THE LAND-ROVER

THE Land Rover
WORKHORSE OF THE WORLD

GRAHAM ROBSON

DAVID & CHARLES
NEWTON ABBOT   LONDON
NORTH POMFRET (VT)   VANCOUVER

ISBN 0 7153 7203 3
Library of Congress Catalog Card Number 76-7294

First published 1976
Second impression 1977
Third impression 1978

Printed in Great Britain
by Redwood Burn Limited, Trowbridge & Esher
for David & Charles (Publishers) Limited
Brunel House   Newton Abbot   Devon

Published in the United States of America
by David & Charles Inc
North Pomfret   Vermont 05053   USA

Published in Canada
by Douglas David & Charles Limited
1875 Welch Street   North Vancouver   BC

To the late Spencer and Maurice Wilks
– for their faith in the Rover company

CONTENTS

LIST OF ILLUSTRATIONS

9

FIGURES

INTRODUCTION
Tom Barton and the Land-Rover

The idea for the Land-Rover came from Maurice and Spencer
Wilks. The former, who was in charge of Rover's technical affairs
at that time (1947), began the planning, but he did not, of course,
design the vehicle on his own – nobody ever does that. There is
always a capable team on hand to express the boss's ideas. Once
the Land-Rover had been launched, however, and post-war cars
needed his attention, Maurice Wilks handed over the Land-
Rover to one of his section leaders, Tom Barton. Barton knew
the project was a stopgap, not expected to last very long, but he
has stayed with it ever since. Whether he likes the title or not, he
is now 'Mr Land-Rover'.

He says of himself:

I wasn't born and bred to be a 4 × 4 designer – far from it! In fact
I started my business apprenticeship in the railway carriage and
wagon works at Wolverton, near Bletchley. I went through every
trade there was in that factory – fitting, turning, painting, everything. At 21 I left there, moved into the Midlands, and started to
work for Metropolitan-Cammell. They made things like Under-
ground trains, other railway stuff, and bodies for buses, which is
how I first got introduced to the motor industry.

When war came along I tried to get into the Army, but everybody said I was in a reserved occupation, and turned me down. I
thought I would cause some trouble and *make* someone call me up,
so I left Metropolitan-Cammell, went to see the Army people
again, and shortly got myself a new job at Rover. The trouble was

that Rover were getting involved in the Whittle jet-engine project, and I found myself in a bigger reserved occupation than ever!

Tom then gave into the inevitable, and immersed himself in turbine engines. For a time he was moved up to Lancashire where Rover were experimenting with gas-turbine prototypes in converted cotton mills, and lived for a time in a rented house with Gordon Bashford, later Rover's designer. But eventually he came back to the Midlands.

In 1943 Rover swopped their gas-turbine work with Rolls-Royce for the rights to build Meteor tank engines at Acocks Green, Birmingham. All had the choice of staying with the turbines, and working for Rolls-Royce, or of staying with Rover.

After the war, my first job when I got away from tank engine work, was to do *the* first left-hand-drive design work for an export Rover. Up to then, quite literally, there had never been one – and Rover had been making cars since 1904! That was in rebuilt offices in Helen Street, Coventry. Gordon Bashford was alongside me, working on future projects, and there was a clay (or perhaps even plasticine) model shop next door. Just before this, Gordon had actually been working in the ballroom at the Chesford Grange Hotel, near Kenilworth.

That was in 1945 and 1946, but shortly afterwards the Solihull works was kitted out and the design office moved into more purpose-built surroundings. Then Barton, and others, were set to work on the Land-Rover.

Maurice Wilks decided that we ought to make a vehicle rather similar to the Jeep. I was one of five section leaders in the design drawing office, and each of us was given the job of designing part of the new device – the Land-Rover. We had a very broad brief, nothing detailed, except that we were asked to make a vehicle similar to the Jeep. My first job was to graft a new transfer box on the back of an existing car gearbox, the P3 box.

We thought about accessories straight away. I also had to think about power take-offs and things – so that the Land-Rover could be useful to the farmer in every way.

Tom Barton's designs must have stemmed from earlier work on railway rolling stock, pioneering gas turbines, and even high-quality vee-12 aero engines, but it is impossible to spot any traces of this in the Land-Rover. Barton himself cannot.

He explains how he became head of the project as follows:

> Once the vehicle was designed, and in production, for a time we only had to look after evolution. We didn't need all the specialist sections; this was considered to be an agricultural vehicle, so the other four section leaders slipped back to urgent private-car work. I was only the railways and gas-turbines expert, so naturally I was left to look after Land-Rovers!

Since then Tom Barton has been loyal to his 4 × 4s through thick and thin, while mergers, reorganisations, financial crises and cutbacks have passed him by. One reason for his pride in his machines is that *'within 12 months* of starting to build Land-Rovers, we were making more of them than the rest of the company was building cars. Since then we've never been beaten . . .'

Tom has been in command of all Land-Rover engineering for a good many years. He has been offered other jobs, but, as he says:

> Once I got started, Land-Rovers became a bit of a drug. We were all enthusiasts from the time we started charging about in the prototypes in all sorts of conditions. I've been all round the world in them, in Army trials, in competitions – I came to love the Land-Rover. I've had opportunities to move on, but always resisted. There were two particular occasions when I was very tempted.
>
> Mr Cullen, who was running Land-Rover development at the time, went off to David Brown Industries to do a 4 × 4 for them, and I was offered his job. What the company didn't know was that Cullen wanted me to go with him! That was in 1954, and I remember that I was given just 24 hours to make up my mind. As you can see, I stayed.
>
> Later, when Colonel Pogmore, my boss, was invited to go to Longbridge to look after the design of the Austin Gipsy, and he asked me if I would go with him, I was very tempted, certainly,

but Rover were quite determined to keep us both. It was soon after that when I was made up to Chief Engineer in 1966.

Before the Ryder Report caused an upheaval in every British Leyland company, Tom Barton controlled more than 100 staff at Solihull, all completely immersed in present and future Land-Rover design. The machine might look the same as it always did, but there have been many important improvements over the years. Even in 1975 Barton's engineers were looking forward to the 1980s, though Tom Barton himself will not be around to see them selling. In 1980 he will reach 65 years of age, and not even a love of the Land-Rover will allow British Leyland to keep him on. But one thing is certain – Tom Barton may have to retire, but the Land-Rover will not.

I

A ROVER FOR THE FARMER

When the Land-Rover was unveiled at the Amsterdam motor
show on 30 April 1948, the world of motoring was astonished.
This was a new model completely out of character for the Rover
company, which was known for making some very dignified cars
for a very respectable clientele.

Rover had been making cars, in Coventry, since 1904. There
had been nothing very remarkable about them before the early
1930s, and more than once the firm might have collapsed for lack
of profits. Rover's saviour was an austere man with legal training
– Spencer Bernau Wilks. He arrived at Rover from Humber-
Hillman in 1929, became General Manager at once, and rose to
Managing Director in January 1933.

When he took over, Spencer Wilks did not like what he saw
and made changes. His new policy for the company was 'Quality
First', and he limited production to a steady and manageable
rate. There was a consistent and dramatic improvement in sales,
and by 1936 he had gained so much prestige for his company that
Rover was asked to join in the Air Ministry's 'shadow factory'
scheme. When Wilks agreed to run the Acocks Green factory in
1936 and the Solihull 'No 2 shadow factory' in 1939, he was auto-
matically committing Rover to an expansive future.

After World War II Britain, heavily in debt, was faced with
the prospect of years of austerity; and industry was directed to
export anything that could earn dollars. Government quotas
and restrictions, however, ensured a shortage of just about every

15

material from which private cars could be made – most particularly sheet steel. The car makers were therefore urged to rationalise, to concentrate on single model ranges, and to 'Export or Die'. Raw materials would be allocated according to a firm's export plans – if it was not prepared to send cars abroad, it would not be allocated supplies for home consumption.

This hurt everybody, changed many a firm, and killed off a few. But Rover, and Spencer Wilks, were in a double quandary: not only did it look as if they would be forced into an unfamiliar field of business, but they would also have to move to a new and unfamiliar factory. Enemy bombing in 1940 had severely damaged the Helen Street factory in Coventry, and Wilks was proposing to transfer all activities to the Solihull buildings once they had been cleared of aero-engine work. But Solihull was big, much bigger than Helen Street, and the labour force would be looking around for work to keep it busy.

Even before the war had started, Wilks had presented an expansion plan to his board of directors, but it had not recommended dramatic increases in output. This was probably just as well, for when Rover applied for Government permission to restart private-car manufacture in 1944, the request was first of all refused, then grudgingly set at a mere 1,100 vehicles!

Wilks was planning on immediate yearly production of 15,000 cars of the pre-war variety, along with another 5,000 of a brand new '6hp' car (the little 700cc 'M' Type), which his brother Maurice was already designing. But with the steel shortage restricting all short-term plans (the 'M' Type, for instance, would have light alloy body and chassis pressings) there was still the problem of keeping his Solihull factory busy.

Early in 1947 it was decided that the 'M' Type would have to be shelved, as it lacked export potential, and Wilks decided that more radical measures would be needed. By now all the immediate post-war car designs had been settled, and there were all manner of possibilities, but the Wilks brothers were not known for making hasty decisions. They were both shrewd business-

The half-million mark was passed in April 1966, and the Board of Directors turned out to honour the achievement. Managing Director William Martin-Hurst is driving and Tom Barton is next to him with A. B. Smith (who retired as Chairman in 1975) to his left. Bernard Jackman is behind A.B., and Peter Wilks on the extreme left of the picture. The Land-Rover is a home-market 88in Series IIA

The two technical chiefs, Maurice Wilks (hand on drawing board) and Robert Boyle, who turned the Land-Rover into a reality

The first Land-Rover prototype at Solihull in 1947. (Note the central driving position, one-piece windscreen, no doors and no hood. Almost everything was an 'extra')

Underside view of the first chassis. Main frame members were strengthened before production. The central power take-off was planned from the outset and appeared on this first chassis

men, and both became aware that there was a worldwide shortage of agricultural vehicles – not through the destruction of existing ones but because farmers were raising their standards. Much of the agricultural world was now ready to mechanise *in toto*.

Spencer and Maurice Wilks became convinced that this market could be the answer to their problem, that they could sell a lot of simple machines at the right sort of price. They had already noted how Harry Ferguson's cheap and effective tractors had started to flood out of Standard's former 'shadow factory' at Banner Lane in Coventry.

The brothers agreed to go their separate ways for a short time, early in 1947, to think things through without outside influences, then meet to discuss their ideas. When they did meet, both had come up with the same scheme, and both listed the same reasons.

Maurice Wilks had provided the spark. He had a house and 250 acres of farming land in North Wales, on the island of Anglesey. To deal with jobs like timber-hauling he had first used a Ford V8-engined half-track, which had not been a success, and then an ex-WD Jeep, which had proved more satisfactory. The story goes that Spencer Wilks asked his brother what he would do when the battered old Jeep finally gave out. 'Buy another one, I suppose', said Maurice. *'There isn't anything else.'*

These words finally convinced them both, and they decided that Rover should be planning something similar. Maurice, as the technical chief, was relishing the prospect of starting, and his brother was soon ready to give the go-ahead.

After the fateful weekend in Anglesey Maurice Wilks arrived in the design office on Monday morning and started planning. He even invented the new name. For better or for worse (and they could not then be certain of success), the Land-Rover was a Wilks brothers' idea.

The final decision was made after a whole series of trials, with the Jeep being thrashed through woodland up and down steep slopes, through water, across the sand dunes, and into the surf along the coastline of North Wales. Yet, at the time, there was no question of the new Land-Rover ever being a long-term project. In the brothers' eyes the new machine would be a useful stopgap, a rugged and practical stopgap, intended to provide the company with some export potential, and the labour force with work. For two or three years, Spencer Wilks thought, the Land-Rover would fill the gap between sales of P3 cars and Solihull's total production capability. It would be a simple little cross-country workhorse – a 'Rover for the farmer'.

It had to be produced, however, without spending a fortune on tools and new capital equipment. Spencer Wilks makes this clear in his own submission to the board of directors. The minute book notes:

> Mr Wilks said that, of the various [sic] alternatives that had been under consideration, he was of the opinion that the all-purpose vehicle on the lines of the Willys-Overland post-war Jeep was the most desirable.
>
> The P3 engine, gearbox and back axle could be used almost in their entirety; little additional jigging and tooling would be necessary, and body dies would not be required . . .

That recommendation was made in September 1947, after design and testing had been going ahead for several months. Farmland surrounding the Solihull works, prudently bought by the Rover company at the beginning of the war, had witnessed all sorts of proving trials by the first sturdy but very simple machines. They were not cars, but neither were they tractors – so what were they?

This was where Rover's genius and Maurice Wilks's logical thinking came into play. Let us expand on Tom Barton's brief, which we have already mentioned:

> Maurice Wilks wanted us to design a vehicle very like the Jeep, but it had to be even more useful to a farmer. That was the point – it was to be a proper farm machine, not just another Jeep. He wanted it to be much more versatile, much more use as a power source. He wanted it to be able to drive things, to have power take-offs *everywhere*, and to have all sorts of bolt-on accessories,

and to be used instead of a tractor at times. It had to be able to do everything!

The early pictures, therefore, of prototype Land-Rovers towing farming equipment – ploughs, harrows and the like – do not illustrate gimmicks. All this and the off-road capability were what Maurice Wilks had asked for.

As we now know, however, the customers themselves decided what their Land-Rovers should do. They wanted to keep a tractor purely for tractor work, it seemed, so the Land-Rover's capabilities in that direction were soon abandoned. In addition, the farmers said, there were a few farm jobs that only a tractor could tackle, two-wheel-drive or not. But for everything else the Land-Rover was ideal.

Even for Rover to engineer and produce such a vehicle was tough enough, but the urgency of the project made it tougher. To fill the factory with work, the usual Rover time-scale of design could not be accepted. Traditionally, of course, Rover designers were simply not used to doing anything in a hurry; they were accustomed to taking two to three years, at least, between first thoughts and first sales. Yet with the Land-Rover the whole process took little more than a year. Maurice Wilks sparked off the project in the spring of 1947, the first prototypes were running by the end of the summer, and the Land-Rover was shown to the world in April 1948. Release was slightly premature, in fact, but the company needed to study first reactions in advance of production, in case any important changes were needed. In the event the changes were few, the vehicle's reception was rapturous, and the first examples were delivered in July. From first thoughts to first deliveries in less than 18 months – could any car-manu-facturer achieve this in the 1970s?

The key to the whole project, logically enough, was the Jeep. Maurice Wilks, of course, had his own Jeep; he admired what it could do for him, and he was certainly not averse to copying any useful features. The designers are adamant that nothing in a production Land-Rover was exactly copied from the Jeep, but they all admit to looking very carefully at every detail, and being encouraged to do so.

The very first prototypes were certainly not pure Rover, by any means. The company's first move was to send a man out to a vast Army surplus dump in the Cotswolds to buy two more Jeeps. Both were immediately torn apart for study, and certain components found their way into the first Land-Rovers, which were definitely half-breeds. To save time, a Jeep chassis was used under the original vehicle, mostly with Rover-designed fittings.

As we have already explained, design went ahead under five section leaders, with Robert Boyle pulling all the right strings and Maurice Wilks in general control. On the mechanical side every possible existing Rover car component – whether an engine, instrument or merely a nut and bolt – had to be used, and the amount of money allocated for new or modified machine tools was minimal. In fact, no money at all was spent on press tools for the bodywork. Panels that could not be formed by simple bending or folding were just not used. Any shape to be found in the skin had to be produced on existing machinery.

The first and only significant miscalculation came at the very beginning, when it was thought that the vehicle would be so light that a Rover 10 engine could be used. This unit found its way into the first prototype, but was found wanting both in power and torque, and was swiftly discarded. Thereafter, and without any dissension, the new 1·6-litre P3 engine was fitted. Although the 1,389cc Rover 10 engine was still in production in 1947, it dated from the 'Wilks New Deal' of 1933 and produced less than 40bhp. There was an alternative enlarged version, the 1,496cc 'Twelve' engine, but this was of similar age. Even though the Land-Rover was only thought of as a stopgap, it was still thought that post-war designs were desirable.

Although the Land-Rover and the Jeep had much in common generally, there was considerable difference in detail. The Jeep, for instance, had a much larger engine, much less space for passengers or goods, and an even more basic specification. The table below compares the two designs. Any design engineer studying this chart would find much to interest him. The basic 'skeletons' – wheelbase dimension, length, wheel tracks, and general engine and transmission layout – are the same, but there are also major differences.

Comparison of Land-Rover (1947) and Jeep

	Land-Rover	Willys Jeep
Wheelbase (in)	80	80
Length (in)	132	133
Width (in)	60	62
Wheel track (in)	50	48
Engine	4-cyl, 1,595cc, ioev, 69·5 × 105mm, 55bhp (gross) at 4,000rpm	4-cyl, 2,199cc, sv, 79·4 × 111·1mm, 60bhp (gross) at 3,600rpm
Transmission	4-speed gearbox, transfer gear, low range (2·52 step down)	3-speed gearbox, transfer gear, low range (1·97 step down)
Axle ratio	5·612 to 1	4·88 to 1
Wheel size	6·00 – 16in	6·00 – 16in
Front suspension	Live axle, half-elliptic springs, telescopic dampers	Live axle, half-elliptic springs, telescopic dampers
Weight (unladen)	2,520lb	2,315lb
Payload	Not quoted on release	'¼-ton', often exceeded – 800lb normal maximum

Considerable constraints were put on the design of the Land-Rover by the Wilks' 'minimum tooling' directive. The chassis frame itself is a case in point. In normal times the main side members would probably have been 'U'-section pressings, the whole thing being jig-built and welded securely together. But in the Land-Rover each side member was of box section (quite modern thinking for 1947) built up very simply from flat steel sheet. Then, as now (even in 1976), the basic load-bearing members of the short-wheelbase Land-Rover are built up, Meccano-style. There are four plates: the side sheets are flat, but curved in outline, whereas the top and bottom sheets start life absolutely flat and straight. These are tacked to the side plates in jigs, before being continuously welded in very low-cost fixtures.

Olaf Poppe, Rover's chief planning engineer, had to devise the production tooling, and his advice on continuous welding re-assured Gordon Bashford that this very basic method could be made to work:

> The original 'four-plate' concept came from Olaf Poppe. So we made up several straight sections from plates, tried out the weld-ing process on each edge, added various cross-members so that we had crude 'ladder' frames, and tried them out for twisting and bending strength. This came immediately after Maurice Wilks had asked me to lay out the original package for a very simple vehicle. It was to be a 'workhorse', and very very basic.
>
> As far as I know this method had never been done before by anyone. Anyway, box sections were still quite new to Rover – all the pre-war chassis had been open channel-section frames.

Bashford also remembers this about the basic concept:

> It was really very much like the Jeep at first. It is no coincidence that the wheelbase and basic dimensions were all repeated in the Land-Rover, as I based my original package around the Jeep. In the very first vehicle we used a lot of Jeep material, and almost automatically that meant we could use the same important dimensions.
>
> The machine I laid out at first had no doors (those were going to be optional extras!), no trim, no hood as such – and it even had a central driving position and steering wheel, with chain drive to a steering box on the appropriate side of the scuttle!
>
> The body, such as it was, would be made out of aluminium. That was partly because of the sheet metal shortage, and partly to give the best possible corrosion protection. It helped a lot when we came to forming the panels, because aluminium alloy is that much easier to work.

This simple form of chassis construction worked well, in spite of some initial misgivings. Another very revealing quote about those early trials is the following:

After the four blank strips were set up in a fixture, we had to tack-weld them. They then went into the automatic welding machine, which had moving heads, and these welded only one side at a time. Now if you can only weld one side of a box section there are some inherent hot-and-cold problems, which tended to make the box twist. Hopefully, when we turned it over and welded the rest it tended to twist itself straight again!

Dropped into this chassis was Rover's own layout for four-wheel drive. The general principles of four-wheel drive are well known, but all designs have their own characteristics. Rover chose to give their machine an alternative set of ultra-low ratios like the Jeep. Special to the Land-Rover was a car-type freewheel in the drive-line to the front axle, which was meant mainly to allow for the front wheels overrunning the rear – even though there were differentials built into front and rear axles. The idea was sound, the engineering competent, and the device already in quantity production for Rover cars, but within a couple of years it would become superfluous.

The use of components from the cars was carried as far as possible. The engine, with its clever and compact arrangement of overhead-inlet-side-exhaust valves, was to all intents and purposes identical with the Rover 60 unit – an engine newly announced to the public only weeks before the Land-Rover; so were the main gearbox, apart from first gear ratio itself, and the internals of the axles.

Thinking about the new project, and drawing the first lines on paper, took very little time, but even with all the urgency in the world behind the prototypes, they took months to build. Once the important transmission parts had appeared, building up the very first vehicle took about six weeks, and the second followed after the same interval. Soon after this, when the design of the chassis had been fixed, a jig was completed and progress became more encouraging. No fewer than twenty-five pre-production (we should probably now call them 'pilot build') Land-Rovers

25

were on test before the spring of 1948. Inside the factory, on
Maurice Wilks's farm, and in the fields surrounding the factory,
the strange and stubby little machines could be seen ploughing,
driving threshing machines, towing rubbish trailers piled high
with animal feedstuffs – doing, it seemed, almost everything.

The new Land-Rover now looked as if it could do all Maurice
Wilks had demanded of it. It would be extremely cheap – £450
was the original asking price for British deliveries – but many
fittings were to be optional extras. There were to be no doors or
sidescreens, no weather protection, no seat cushions for passengers, no heater, no spare tyre (though at least there would be a
spare wheel rim), and no starting handle. On the other hand
there would be left- or right-hand-steering versions, the original
idea of central steering having been discarded.

Whether or not the planners could get it ready for production by
the end of April 1948 the Land-Rover had to be announced then.
The Amsterdam motor show was an important world-market
occasion, and the Wilks brothers wanted their brainchild to be
there. Even though it was not ready to be sold at once, it came
along only just in time.

Rover had set up their export department in 1945, but their
private cars were simply not of a type likely to appeal in far-flung
territories. Because of the government's 'no exports – no steel'
edict, the situation was critical. Car production had crawled ahead
slowly from thirty cars a week in January 1946 to 200 a week by
the end of that year. Power shortages in 1947 took their toll, and
even for 1948 it looked as if only 3,000 cars would be built in the
entire year. It was not as if Rover lacked orders – in Britain they
were faced with tens of thousands – but orders for exports were
lacking.

Perhaps the Land-Rover would make up the deficiency. The
Wilks brothers thought they might start by selling 100 Land-
Rovers every week – about 5,000 in the first full year – but the
officials in Whitehall had other ideas. In December 1947 an un-
official allocation of steel had been promised, sufficient only to

produce 1,000 Land-Rovers in a year! It was in an atmosphere of doubt and uncertainty, therefore, that the Land-Rover was launched upon the world. Its designers and planners had done their best. Now it was up to the customers.

2

THE DEVELOPMENT OF
THE LAND-ROVER

The trouble with starting a trend, or with inventing any new type of machine, is that your name goes into instant folklore. For years any vacuum cleaner was a Hoover, and any ballpoint pen was a Biro. The Land-Rover had the same problem. Any competitive machine, whatever its nationality, was still a 'Land-Rover'. Only the Jeep maintained its individuality.

From time to time other makers have tried to beat Solihull at its own game, mostly without success. One particularly determined attempt by BMC – with the Austin Gipsy – provoked Rover in 1958 into stating: 'When better Land-Rovers are made, the Rover company will make them.' What Rover were really saying was 'Beware of all imitations'.

For a generation the relation between the Land-Rover and its customers had been close. The Land-Rover was what they wanted, and many of them were convinced that they had invented the need for it personally. The slogan used by Vauxhall in Britain for their new Chevette car – 'It's whatever you want it to be' – applies equally well to the Land-Rover. The Wilks brothers may have had the original idea, but it was the customers who guided its development thereafter.

In 1948 its reception was ecstatic. Nobody, apparently, had a bad word for it. The motoring press loved it, the agricultural press loved it – and the customers flocked round to place their orders.

In a matter of weeks Rover had a big problem on their hands – how to make and deliver enough Land-Rovers. It was already clear that 100 Land-Rovers a week would not satisfy the demand, and compared with the doldrums of private car sales at the time, this was a magnificent start. The interest, indeed, has continued, and the Solihull company has been faced with a similar dilemma for nearly 30 years.

The reason for the rush of orders soon made itself plain. Whereas Maurice Wilks had foreseen the Land-Rover as the far-mer's friend, the customers saw it as everybody's friend. The farmers certainly ordered thousands of the new machines, but so did the contractors, the builders, the police forces, the motoring organisations, the military, the explorers and so on.

The first deliveries were made in July 1948. (Incidentally, the first-ever production Land-Rover now lives in happy and proud retirement at Solihull.) At first the customers had little choice, except in the matter of colour, and which of the 'essential extras' were fitted before delivery.

The 80in wheelbase Land-Rovers on the first production line differed very little from the vehicle shown to the press, but they were vastly different from the first prototypes of 1947. The central steering idea had, of course, gone; there was now pro-vision for passenger doors, though, theoretically, these were still 'extras'; the rear bodywork had even less form than at first; and there was now less shaping in the front wings and the bumper. Under the skin the chassis was already more solid than that first considered, and any visible affinity to the Jeep had disappeared.

Rover are now proud to describe the Land-Rover as 'the world's most versatile vehicle', mainly because there are so many different possible versions. In the summer of 1948 this could only have been a dream. A customer could have his 80in machine, with or without a canvas hood, and with or without such valuable extras as the power take-offs.

The expansion process started within months, and in October a chubby and rather attractive station-wagon version arrived.

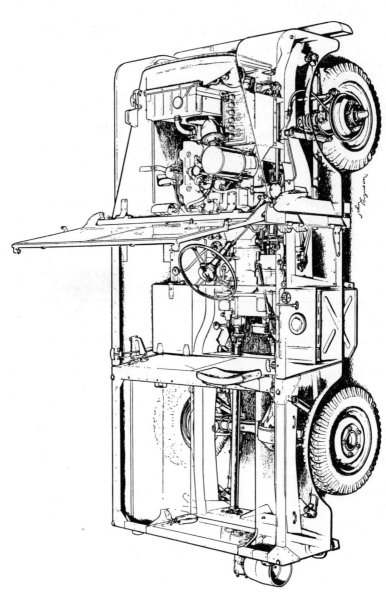

FIG 1 The first production Land-Rover laid bare. The *Autocar*'s cutaway drawing shows the (now-classic) layout, and how easily the P3 engine fitted under the stubby bonnet

Mechanically it was unchanged, but the bodywork was mainly new. Not only were the doors standard, but they had winding glass, there were four inward-facing seats in the rear, and access was through a two-piece (up and down folding arrangement) tailgate. The panelling, of course, was in light-alloy Birmabright and weather protection was complete. The price, however, at £959 inclusive of British purchase tax, was more than twice that of the basic vehicle, which explains why it was not a success at that time.

British tax law relating to cars, agricultural vehicles, four-wheel drive commercial machinery, and dual-purpose transport, was extremely complex in those days, and the versatility of the Land-Rover made it difficult to categorise. Private cars had to pay purchase tax, while commercial vehicles did not. A station wagon had to pay tax because the authorities considered it as a 'private and domestic' machine, unless it had twelve seats – when they pronounced it a bus, and not subject to tax any more!

In Britain at the end of the 1940s petrol was still strictly rationed, and to differentiate between private and business motoring there were two grades of 'Pool' petrol; that for commercial use, and supplied against appropriate ration coupons, was tinged with a red dye. But how should a farmer fuel his Land-Rover? After all, he had many legitimate business uses for it, on and off the roads, and unless he was very rich, he might also have to use it as his 'car'. There were several prosecutions over that question.

The Land-Rover's status took years to sort out. The question of seating capacity in estate cars was eventually resolved by rule of thumb, the seven-seat and ten-seat wagons being subjected to purchase tax, and the twelve-seat wagon not.

Whether the Land-Rover was a commercial vehicle or not mattered greatly to British customers. Open-road speed-limit law in the 1940s and 1950s restricted a commercial vehicle to 30mph. As the Land-Rover was taxed (or rather *not* taxed) as a commercial vehicle, should it not comply with all the laws governing that category? The police thought so for years, and it needed one celebrated test case in 1956 to change their viewpoint.

Mr C. Kidson of Wareham in Dorset, who had been fined £3 for exceeding 30mph on the open road in his Land-Rover, appealed and won his case. Lord Chief Justice Goddard ruled that a Land-Rover was not a commercial but a dual-purpose vehicle, within the meaning of the Act. According to the judgement, 'Mr Kidson was driving a modern type of Land-Rover which was more like a truck. It could carry passengers in the cab and goods in the back, but it was capable of being used as a four-wheel drive vehicle, and this brought it within the definition of a dual-purpose vehicle. It was not, therefore, subject to the 30mph limit on goods vehicles.' The Rover company, and thousands of its customers, breathed a collective sigh of relief.

This court ruling illustrates how the Land-Rover had cut a swathe through the accepted definitions of many motoring classes, to set up its own category. None of the legal rulings of these early days have ever been reversed. In Britain today, of course, the supersession of Purchase Tax by Value Added Tax has caused wholesale changes in Land-Rover prices, since each and every one, whether open, closed, station wagon, van or pick-up, is now subject to VAT. Some versions also attract Car Tax.

Spencer Wilks soon changed his plans for the Land-Rover. Within months he knew that his 'stopgap' philosophy would have to be discarded, and by 1950, even after the arrival of the newly styled P4 private car range, he had realised that the Land-Rover had taken over at Solihull. The Land-Rover was to dictate Rover's fortunes for the next generation. It was already bigger than its parent, and the parent had come to rely on it. The 'stopgap' of 1948 was already the 'mainstay' of the 1950s.

Production figures illustrate the Land-Rover's success. During the financial year 1947-8 only forty-eight Land-Rovers were delivered, but in the next 12 months a carefully budgeted 8,000 were made, and in the following year no fewer than 16,085 rolled out of the gates. Demand thereafter caused rising sales almost every year, with only the British 'credit squeeze' year of 1956-7 providing the exception to the rising totals.

Within a year, the Land-Rover had overtaken Rover car sales – in 1948-9 these were 5,709, and in 1949-50 only 3,944 units. By 1951, with the Land-Rover still less than three years old in the market place, it was outselling the cars by two to one.

Maurice Wilks had had no intention of widening the scope of his Land-Rover when it had just been considered a stopgap, but once it had become a runaway success his intention changed. Since 1950 the process of change has been continuous, and the objectives consistent. Any change would only be accepted if it made the Land-Rover more versatile, more rugged, more reliable – and if it increased sales.

By now the conventionally laid-out machines have been sold in five different wheelbase lengths, with the option of four different petrol and two diesel engines. Body variations have proliferated – open, truck cab, van body, estate/wagon body – with special fittings in profusion and, eventually, a large number of approved conversions by outside specialists. Consideration of specialised types of Land-Rover, such as those that float, fight fires, support airline steps, clean lighting standards, and the like, is deferred to Chapter 4.

The Land-Rover's most active development period, when it was a dull year that passed without some major improvement, was the 1950s. Apart from the addition of the six-cylinder engine option in 1967, the 1960s and 1970s to date have been relatively calm, though sales have continued to roar ahead, with still higher sales being forecast for 1975-6.

The years 1950 and 1951 saw the first widening of the range, and some features were slipped into production without much trumpet-blowing. The station wagon, killed (as we have seen) by its high cost, was withdrawn, and at about the same time the freewheel feature was discontinued. This had been a technically sound feature, necessary according to the mechanics of the four-wheel drive layout, but in service it was found not to be

33

essential. Before the station wagon disappeared, a metal 'van' became available, and because this had no side windows, it attracted no tax.

At about the same time the headlamps came out of their hiding place behind the mesh grille, and the original modest facia was replaced by a panel displaying larger and more appropriately styled instruments. But there was still no leaning towards comfort – glove lockers and padding were still many years ahead.

The original Land-Rover proved itself so useful that most customers persisted in overloading it – then complained that the engine was not powerful enough. The company listened, politely demurred at first, but eventually bowed to the inevitable. Not even Volkswagen, with the Beetle, could resist the temptation to make a world-beating machine a little quicker. From the beginning of 1952 all Land-Rovers were to be built with a more powerful 1,997cc engine – one that was externally the same as the original 1,595cc unit. Neither engine at that moment was shared with a car: the obsolete engine had last appeared in the 1949 P3 saloon, while the new unit would not be found in a P4 (in its '60' guise) until the end of 1953.

Owners of Land-Rovers went on overloading their machines, but now started to wish for more space in the back. The original vehicle, admittedly, was only 11ft long, with little more than 4ft of loading platform behind the seats. In 1952, therefore, Tom Barton's team was encouraged to do something about this for the 1954 models, and in two different ways.

On the short chassis they added 9in to the length of the loading platform. While the engine, transmissions and passenger compartment kept their distance from the front axle, the main chassis members were reshaped and extended. The rear axle was effectively moved back 6in (thus increasing the wheelbase to 86in), and an extra 3in were added to the extreme tail. It was, at first glance, a simple thing to do, but when all the implications of new chassis members, new bodywork, new propeller shaft, new rear springs, new exhaust pipes and a mass of other detail were

The production format develops. The two-piece windscreen appears on this car and the unique chassis welding is in evidence

The 80in 'rolling chassis'. Note the tiny instruments and simple facia

Ernest Marples and his bride-to-be with the special-bodied Land-Rover caravan they used for honeymoon transport

A famous Land-Rover, the first production vehicle, now in retirement in Solihull

considered, the 86in Land-Rover was virtually a new model.

That was only the start, for Maurice Wilks had decided to silence his customers' complaints once and for all. He would also offer them a Land-Rover with a completely new chassis frame and a wheelbase of no less than 107in! This was 21in longer than the new Regular Land-Rover and 27in longer than the original 1947 layout, and guaranteed an extra 41in of loading platform length behind the driver's seat. There should be no more complaints on that score.

Nor were there any. Solihull's production planners had problems in catering for alternative frames in an already busy chassis-building section, but as there was still some common use of cross-members and bracketry, they soon came to terms with it.

At the same time the opportunity was taken to include a 107in ten-seat station wagon in the range, which added to the number of choices available to the Land-Rover customer. The basic machine did not look much different, for all that, certainly not from the front. It still had the same bluff nose, the no-nonsense bumpers, and, as often as not, the spare wheel mounted atop the bonnet pressing.

A couple of years later came a change that made little sense at the time. The wheelbase, and therefore the length of both versions, went up by a mere 2in. The '86' and '107' models became '88s' and '109s'. Rover were apparently far too busy building and selling 500 Land-Rovers every week to explain a mere 2in, but in fact they were added to make way for an important new feature, about which the company's lips were sealed. That extra couple of inches had found their way into the wheelbases, slotted neatly into the space between the gearbox and the line of the front axle. The existing engine needed no more elbow room – but a new diesel did!

This, then, was the key to Rover's next big step forward. Nobody had to convince them that a diesel engine option was desirable. In some foreign countries it was much easier to find diesel fuel than even low-grade petrol, and the operator with large trucks, excavation equipment and stationary engines probably wanted to use nothing else. To such people a petrol-powered Land-Rover was only an embarrassment.

Rover's only problem up to then was that such things took time, extra capital, new transfer-line tooling, and a great deal of painstaking development. Of course, they could have bought diesel engines 'off the shelf', but that was not Rover's way of doing things. If the proposed merger with Standard had materialised in 1954, the Ferguson-type diesel might have been made to fit; but in the event, a Rover-designed diesel was the only answer.

The optional diesel was released in June 1957, but it had been under development for at least 4 years before that. The new unit, of 2,052cc, was unique to the Land-Rover, and completely different from the existing four-cylinder petrol engine. General dimensions were larger and more robust, which explains the extra wheelbase needed; and conventional overhead valve gear was fitted, since Rover's own unique valve gear was not at all suitable for the very high compression needed by a diesel version.

While these engineering developments were going on, sales and publicity successes mounted fast. King George VI was an early customer, and he was soon joined by Sir Winston Churchill. The British Army (see Chapter 3) took the Land-Rover to its heart, and adopted it as its standard ¼-ton 4 × 4 vehicle in 1956. For Royal visits to military parades, tribal gatherings, horse shows, or other off-road occasions, a specially converted Land-Rover soon became standard. The BBC tried very hard not to name the Queen's transport, but after dabbling briefly with the phrase 'field car' gave in to the inevitable.

With the diesel-engined option firmly launched, even more significant developments came under consideration. After 10 successful years, and nearly a quarter of a million deliveries, the company decided to change the Land-Rover's looks! To those who cherished the absolute simplicity of the machine, this must have sounded like sacrilege, but there were good reasons. Land-Rover and Jeep copies were now beginning to appear from other countries – some successful enough in some territories – and most of the manufacturers were giving their versions a semblance of style and comfort. Once Rover heard of BMC's intentions at Longbridge, they knew that functional pride would have to be submerged, if only slightly.

For 1958, then, the Land-Rover would have to be restyled – or rather, it would have to be styled, for the very first time. Having delivered himself of a splendid company flagship in the big P5 3-litre saloon, David Bache was asked to deal with the exact opposite – the reshaping of the Land-Rover.

He admits that it was not easy. He said:

> The Land-Rover was so absolutely right already that we couldn't just dash in with some obvious improvements. It just wasn't the sort of machine that cried out for decoration, and because of the nature of its work there was no point in putting a lot of delicate shape into the panels.

Changes were subtle and practical. Two fresh air intakes were added to the base of the windscreen panel, a little more form was given to the bonnet pressing, and some crowning was added to the wings, doors and side panels, so that the new Land-Rover would be less aggressively slab-sided. One tiny touch of modesty, and an obvious recognition point from the side, was that sill panels were added under wings and doors to hide the chassis and suspension. Not an ounce of sheer practicality was lost.

The one big technical change was that the old P3-based 1,997cc engine was dropped, to be replaced by a new 2,286cc petrol engine, designed around the same dimensions as the diesel unit and made on the same production lines. Simple overhead valve gear was chosen, and the new engine produced 70bhp at 4,250rpm, with 124lb ft of torque at 2,500rpm. Wheel tracks were increased by 1·5in, and the rather stately turning circle cut by more than 3ft. There was even a touch of softness in the new vehicle – door panels were optionally trimmed, the seat became adjustable on 109in models, and floor carpet was an extra.

Because of inflation, prices of the new Series II Land-Rovers were rather higher than before. Compared with the £450 of 1948, the regular 88in wheelbase machine cost £640 and the 109in version £730. Fitting a diesel brought big benefits in economy, but added £90 to the purchase price.

This, then, was the 'better Land-Rover', whose coming the company had first delicately hinted at a few weeks previously. It was certainly no disappointment to its clientele, who queued up in ever-increasing numbers; in the first full year of Series II production more than 28,000 were built, and a year later this shot up to more than 34,000.

Pressures on factory space were now intense. The Perry Barr factory, which now builds front and rear axles for Land-Rovers and Range Rovers, had been bought in 1952, and the Percy Road building (Land-Rover gearboxes) had been annexed a couple of years later. This helped a bit, but with Land-Rover production now more than 700 a week and still rising, new models in the offing, and all manner of ambitions in mind, Rover management needed a bigger factory.

The answer was a new Rover building site near Cardiff, approved in principle at the end of the 1950s, where a factory was due to be finished by 1963; and the bringing into use of a big new block at Solihull, near the north boundary. The Cardiff factory might have taken over the building of Land-Rovers, or the brand new P6 Rover 2000 car, but in the end it became an important centre for building car transmissions and suspensions, and the home of the complete Service and Spare Parts divisions.

The layout, specifications, and options available to a Land-Rover customer might now be looking settled, but in the next few years there were to be yet more improvements. Only 3 years or so after release of the Series IIs the vehicles were uprated yet again, to become Series IIAs. The main mechanical change was the enlargement of the optional four-cylinder diesel engine from

2,052cc to 2,286cc, so that it had the same bore and stroke as the petrol engine.

There would be yet another engine option to come, and this would represent a clear change of direction in the conception of the Land-Rover. It would mark the first of the moves towards 'civilised' Land-Rovers, a process that is still evolving (see Chapter 5).

During the 1960s spotters outside the gates of the Solihull factory would have noted stranger and ever stranger vehicles being tested. Apart from the military Land-Rovers, some of which were much more specialised than one might imagine, Tom Barton's engineers were turning to new breeds altogether. In 1960, for instance, the first of the ungainly but effective forward-control Land-Rovers took the road. By sticking to the Land-Rover name, without embellishment, for so long, Rover have often hidden the very real improvements and changes present in a new model. The 30cwt-payload forward-control machine was a rather special animal, and the company should have said so.

The new machine came about because Rover directors wanted to extend their range. Even the 109in Series II vehicle was limited to a 15cwt payload, so they requested the design team to produce one with a 30cwt payload and more spacious loading area. It would be called a Land-Rover, but it would really be a light truck. It would not be offered in dozens of versions, as it was intended to have one standardised three-seater truck cab. Naturally it would inherit all the Land-Rover's cross-country capabilities, and most of the existing mechanical options, such as power take-offs and winches, would be available.

When the public first saw the new device, at the 1962 Commercial Motor Show, it brought many of them up with a start. A conventional Land-Rover, especially in station-wagon form, was almost recognisably a car – but this was something else. It had a sturdy bluff-fronted cab, carrying the familiar badge, and many recognisable components, including the screen and doors; but it was well up in the air, on a 109in wheelbase chassis and 9·00 – 16in

41

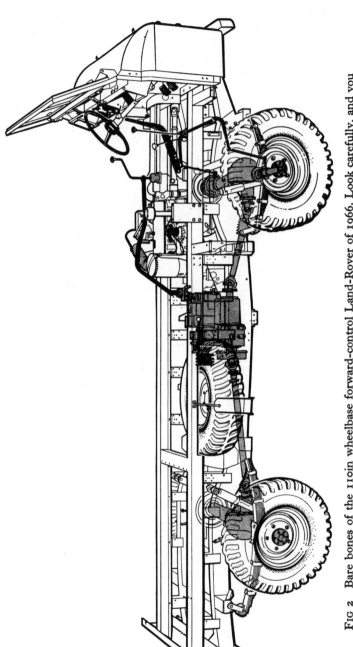

FIG 2   Bare bones of the 110in wheelbase forward-control Land-Rover of 1966. Look carefully, and you will see the entire normal-control 109in chassis as its basis

wheels, with an impressive minimum ground clearance of 10in. The engine was the usual 2,286cc petrol unit, and the transmissions all looked like the usual Land-Rover items.

That wheelbase dimension was significant. To save time and capital tooling expense, and to preserve interchangeability, Tom Barton's men had retained the existing long Land-Rover chassis frame, and mated to it a complete extra top frame to support the cab and seats, and the drop-side body. Rover advertising claimed that the new model used at least 75 per cent of existing components.

It was quite a lot heavier than the normal 109in Land-Rover, and perhaps its lack of alternative bodies weighed against it. There could have been few complaints at the price of £1,015 tax-free. Whatever the reasons, sales of the new machine were somewhat disappointing. To rectify this, in 1966 the vehicle was thoroughly revised. It was decided to change the range of engines: the 2,286cc diesel and for the first time a six-cylinder 2,625cc petrol engine (inherited from the Rover P5 passenger car) were offered, while the original 2,286cc petrol engine was kept for export only. To get the 'six' into the chassis, the front axle was moved forward by an inch, and the new wheelbase became 110in. This, incidentally, is the longest yet seen on a Solihull-built Land-Rover. To aid stability, the axles were widened, the tracks going up, by 4in, to 57·5in. One odd thing was that the original forward-control machine had used the 2,286cc petrol engine, but this engine was the only variant not now available in Britain.

This machine was thought an odd-looker by some, but it could not compete in oddity with some of the 'specials' dreamed up by outside teams of engineers. Even the military Land-Rovers looked very singular indeed, though they only illustrated the versatility of a concept that had not yet reached the limit of its capabilities. Small wonder that the Land-Rover was a magnet to attract envious glances from other car manufacturers.

3

MILITARY MACHINERY

The British army, the Royal Navy and the RAF all use Land-Rovers, as do foreign servicemen, all over the world. In Africa, the Middle East, the NATO countries, the Far East and Asia, and in many tiny states there are Land-Rovers in battledress. Along with the Jeep the Land-Rover is one of the most widely used military vehicles of all time.

One surprising thing about Land-Rovers for military use is that they may be nearly standard, and another is the number of minor differences specified by various customers. Like the civil versions on which they are all based, there are nearly as many different specifications as there are buyers.

A military Land-Rover is nothing if not versatile. It can be made to swim or to go by aeroplane. It must function in sandy, wet, arctic or tropical conditions, and there are dozens of conversions to make a military Land-Rover even more special. Not that such things are carried out at Solihull, though one can be sure that factory engineers know about them.

Land-Rovers can be dropped by parachute, sometimes with de-mountable bodies for extra convenience, or fitted with armour plate. They may act as gun carriages, mobile radar stations, air-traffic control vehicles, scout cars, ambulances – even riot control and prison vans. Probably, indeed, they have many more secret uses, which the world's defence and security forces are not yet willing to reveal.

Land-Rovers first appeared in battledress in the late 1940s,

The first Land-Rover production line at Solihull in 1948. Building work is still going on to the right of the picture

The Land-Rover station wagon of 1948–51. Short, capacious and practical – but far too expensive

Proving the product. The safe operating angle might have been 30 degrees, but all Land-Rovers can tilt this far without capsizing. An 80in model under test for stability

A more civilised truck cab with one-piece screen and doors with wind-up windows

There was very little cargo space in those early vehicles

The 100,000th example, built in the autumn of 1954. It was an 86in Land-Rover, destined for export

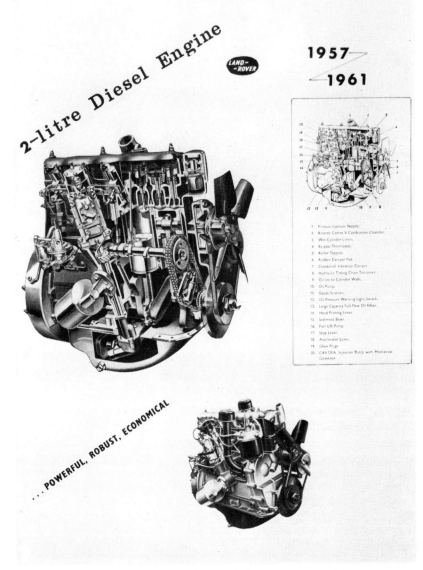

2-litre Diesel Engine

LAND-ROVER

1957
1961

... POWERFUL, ROBUST, ECONOMICAL

1. Pintaux Injection Nozzle.
2. Ricardo Comet V Combustion Chamber.
3. Wet Cylinder Liners.
4. By-pass Thermostat.
5. Roller Tappets.
6. Rubber Damper Pad.
7. Crankshaft Vibration Damper.
8. Hydraulic Timing Chain Tensioner.
9. Oil Jets to Cylinder Walls.
10. Oil Pump.
11. Gauze Strainer.
12. Oil Pressure Warning Light Switch.
13. Large Capacity Full-Flow Oil Filter.
14. Hand Priming Lever.
15. Sediment Bowl.
16. Fuel Lift Pump.
17. Stop Lever.
18. Accelerator Lever.
19. Glow Plugs.
20. CAV DPA Injection Pump with Mechanical Governor.

Rover's first diesel engine for the Land-Rover, announced in 1957. In enlarged 2,286cc form, it is still in use in 1976

superficially as logical successors to the US-built Jeeps. The Land-Rover, however, as it did so often in other markets and other conditions, eventually carved out its own niche and defined its own unique abilities.

The military story really begins in 1943, when Jeeps were being supplied to British forces in large numbers. At the time it could not be assumed that an early end to the fighting was in prospect, and to save precious dollars the government decided that a new design of 5-cwt payload 'British Jeep' should be developed.

The motor industry was invited to submit designs to the FVDE, for a vehicle that was then intended to be an amalgam of components from various British companies. It might, for instance, have had a Rootes engine, Austin transmission, and another company's structure. There would be one chosen assembly contractor, but the other companies would supply major components. Eventually Nuffield Mechanisations Ltd took on the design job, named Project FV1800, and by the end of the European war it had reached the prototype stage. The engine was the Issigonis-inspired flat-four Morris unit, and the whole vehicle was much smaller than it would finally become.

Although Rover were steady and respected suppliers of military machinery, they never took part in the FV1800 schemes. Spencer Wilks, who might have been tempted to tender for work if a large number of 'British Jeeps' had been contemplated, was not interested in the small quantities proposed.

The FV1800, later to be named 'Champ' in civil guise, was put into production by Austin soon after the formation of BMC, but development had taken so long that the Land-Rover was already being sold in large quantities. Unlike Solihull's product, the FV1800 was not a simple design. Production examples used the four-cylinder B40 2·8 litre Rolls-Royce engine, and had a five-speed gearbox, but there was no elegant drive-splitting transfer gearbox behind it. Instead, drive was taken direct to the rear differential, and front-wheel drive could be engaged via a long propeller shaft connecting this with the front axle. There was a

pure cruciform chassis frame, and all-round independent suspension by wishbones and torsion bars. One feature of the
FV1800 was its splendid agility, due partly to the large wheel
movements of at least 10in.

Although the vehicle was quite compact, with an 84in wheelbase and a 12ft overall length, it was also heavy (about 3,500lb)
and used a lot of fuel. By the time Austin put the FV1800 into
production in 1952 they had secured permission to sell it to
civilian customers overseas, and there was the option of a de-
tuned Austin A90 engine. Its versatility in service, military or
civil, was limited in that it could not be used for carrying freight,
and it had only a single, very spartan, body style, open to the
elements apart from a canvas hood and plastic sidescreens.

In the meantime, the Land-Rover had rapidly made a name
for itself. Even as early as 1947, FVDE had looked at the very
first prototypes, and taken delivery of three from the initial
batch (one with left-hand drive) for initial assessment. It very
quickly became clear that, although the Solihull-built machine
was not precisely tailored to any forces' requirement, it had
attractions of its own.

The first trial order, for 50 Series I Land-Rovers, was placed
soon after announcement, and these entered army service in
1949. This order was soon followed by a more substantial order
for several hundred, and since then purchases of the Land-
Rover by the army have gone from strength to strength. Whereas
the purpose-built FV1800 was only intended for a 'once-and-
for-all' bulk order, Land-Rover contracts were soon being placed
regularly.

The two vehicles arrived, operationally, at a time of transition.
This was the beginning of the era, for British forces, of the
'General Service' vehicle – one having a large number of commercially available parts, but not necessarily being quite as agile
or mobile as a fighting vehicle like the FV1800 was. Because of its
specialised nature, the FV1800·Champ was also much more expensive than the Land-Rover, which was almost exactly what the

British forces were looking for. It was very closely based on a regular-production vehicle, it had performance equal to the old Willys Jeep, and the same impressive mobility.

Although the army was already committed to its FV1800s, and had to limit its orders for Land-Rovers at first, it soon found that the two machines dovetailed neatly. The FV1800 was specifically developed as a 'high mobility' combat vehicle, one that could take its place alongside fighting troops at the front, whereas the Land-Rover was ideally suited to operations behind the lines, as an off-road all-purpose vehicle for HQ and other arms of the service needing transport with agility and good performance.

Land-Rovers first went into action in Korea, between 1950 and 1952, where it soon became clear that they were more versatile than the FV1800s, even though they were significantly less powerful and less specialised. Whereas Champs had four basic seats, and were suitable for commanders and their staff, Land-Rovers could be used for carrying people, stores, or any combination of the two. Then, as now, they had no armour protection of any kind, being, in military terms, 'soft skin' vehicles.

Its early experiences with Land-Rovers persuaded the army that they needed more power and more space. The original 5cwt rating meant load plus driver, but in wartime the willing Land-Rover was often asked to do more than it had been developed to do. Wheelbase increases and engine improvements (see Chapter 2) were greatly appreciated, and by the mid-1950s, when FV1800 production had ceased, the Land-Rover was its equal in almost every way. The 'in the firing line' capability of a small 4 × 4 was not seriously missed, as the army was already turning towards armoured personnel carriers for that role. Finally, in 1956, came the accolade – the British Army adopted the Land-Rover as its standard ¼-ton 4 × 4 'Forward Area' vehicle.

Twenty years on, the Land-Rover retains its proud position. As Rover's specialist in government sales, Lt-Col Pender-Cudlip, told me: 'Nowadays it is nice to know that the Land-Rover is the most numerous *and* the most permanent vehicle in our Army. In theory they have to reappraise it every five years, but now, of course, there is no serious competitor in this country.'

MVEE make the point that, nowadays, there are really two types of military Land-Rover. One is rather specialised, with an impressive amount of additional equipment built on at Solihull, while the other is very close to the normal 'civilian' specification. For foreign countries either specification can be duplicated, of course. One very useful variant is the FFR (Fitted For Radio) Land-Rover, with a complete 24-volt electrical system.

One military claim is that, over the years, the army's increasingly arduous requirements have tended to improve everyday Land-Rovers. What was the 'military special' of a few years ago might now be the standard commercial vehicle of today. One good example of this claim was the very tough gun-towing requirement for the 109in Land-Rover. All sorts of non-standard loads were inflicted on the rear axle, and more and more modifications were made to it, until eventually a complete Salisbury component was fitted in its place. This later found its way into all production machines. 'Ambulance suspension' also, which was requested to give an improved ride, was soon standardised on all military 109in Land-Rovers, and has now been added to the basic version too.

The Land-Rover's versatility, and its cargo-carrying abilities, filled another gap in the services' line-up. After the war production of Austin Ten 'Tilly' pick-ups dried up, and 15cwt production also ceased when the manufacturers found that it was an unpopular civilian size. The services, therefore, turned to the Land-Rover, when it matured, and it immediately made itself very popular.

Radio equipment, machinery, armaments and kit can be carried over very difficult terrain, in addition to troops and their personal weapons. The need for troops to jump out quickly in an emergency explains why there are so few station-wagon versions in active service; the army tends to go for armoured vehicles if total enclosure is needed.

Army officers admit that they load their vehicles to the limit, and in service and in emergencies overload them for short journeys. That is one reason why the British have not ordered many diesel-engined Land-Rovers. At one time the British, along with other NATO countries, were searching for the ideal of the multi-fuel engine, but this has now been dropped. Adequate supplies of any fuel, even in the midst of conflict, can now be guaranteed.

The important figure in any vehicle specification is its permissible gross weight. Since a diesel engine is unavoidably rather heavier than its petrol equivalent, and somewhat less lively, it finds no place in the British Land-Rover line-up, except in certain areas where spark ignition engines are taboo for safety reasons.

All of which makes it difficult to understand why another NATO country, in 1975, ordered large numbers of Land-Rovers with diesel power!

In addition, diesel-powered machinery really begins to pay for itself by intensive use, and over high annual mileages, and neither of these conditions are fulfilled by the forces of the 1970s.

If you were to witness the army on the move nowadays, the variety of transport might surprise you. Naturally, the familiar Land-Rovers would be much in evidence, but a number of other vehicles might look vaguely familiar also. They would have Land-Rover badges, but why were they so very different from civilian machines?

The story, as with many military developments, goes back many years, before World War II. Field commanders then often wished that their infantrymen could be equipped with lightweight transport, dropped by parachute. Majors who eventually became generals rarely forgot this dream, and sponsored a series of abortive experiments. First of all attempts were made to parachute Jeeps from the air, but the vehicle usually ended up looking rather like a khaki banana, and then a variety of eccentric but entirely unsuccessful prototype machines were put up by the

53

motor industry, and these generally suffered the same fate.

In the 1950s and 1960s, however, the army decided that such a project should be studied seriously, and it was agreed that a purpose-built machine ought to be evolved. Until the 1950s, certainly, the only way that a car could be parachuted to the battle front was from a cargo plane, but this imposed severe strains on a vehicle's chassis, even when mounted on shock-absorbing platforms.

Only after the later-generation large helicopters became available did the concept of carrying a car to a forward point and gently lowering it to the ground become feasible. We tend to forget that the so-useful helicopter was not developed in any form for use until after World War II.

The idea of a lightweight vehicle for the PBI was ever-present, and once the splendid Westland Wessex helicopter entered service, the dream might at least become a reality. Experiments showed that the Wessex might carry troops or equipment inside its belly *and* find enough lifting power to carry at least 2,500lb underneath on its hook.

That figure, of 2,500lb, was a difficult challenge, and too ambitious for existing Land-Rovers to meet. Other manufacturers were invited to meet the challenge, and BMC, with their lightweight Mini-Moke (Mini-based) and Austin Ant (1100-based, but with four-wheel drive) tried hard. Technically, both could have been useful, but they lacked ground clearance, and would in any case have cost as much as a Land-Rover. It looked like stalemate – the services could not relax their weight limits by much, but manufacturers could not get down to it without losing strength.

At Solihull, Rover took another look at the problem. The 109in Land-Rover could never be reduced to such a weight, but the 88in might. Suppose one discarded the screen, the doors, the rear body, most of the seats and the wings? But on an existing machine these were all 'built-in' on the production line.

The solution was radical, but effective. Tom Barton's en-

54

gineers would leave the chassis and mechanicals alone, but propose a completely different body. What resulted was an oddly styled body, with as much as possible designed to be taken off. Ready for helicopter transport, the Land-Rover that remains is still completely drivable, useful, and battleworthy, even though it looks austere and curiously naked.

In style and construction this very special little vehicle is a throwback to the original Jeeps, or even to the prototype Land-Rovers. Frontal 'styling' is certainly individual, and apart from the bonnet panel itself there are no shaped panels of any sort. Even the bonnet relies only on simple folds, the sort that any technician could repair after minor damage.

Even with all this effort, the stripped-out machine could not quite get down to the terribly demanding limit. However, at 2,66olb it was a very useful compromise, and it was found that the Wessex helicopter with a correspondingly reduced internal load could airlift that much. The MVEE then gave it a new name – the '$\frac{1}{2}$-ton' – which caused some confusion. It certainly carried no more than earlier 88in vehicles, but the '$\frac{1}{2}$-ton' in this case referred to a total payload.

There have been two different-looking versions of the '$\frac{1}{2}$-ton'. At first, from 1966 to 1971, the headlamps were carried in the front grille panel, with side lamps and indicators in the front 'wings'. Since then the machine has been built with all forward facing lights in a modified front wing, ahead of the front wheels.

Named 'Rover 1' by the military, the lightweight Land-Rover is taking over from existing normal-bodied 'Rover Mk 10' 88in machines, which are now obsolescent. Indeed, overseas orders from military and security forces for 88in Land-Rovers are usually filled by the starkly equipped and immediately recognisable '$\frac{1}{2}$-ton' version.

By the early 1970s, the Land-Rover was almost a universal tool of security forces. Civil Land-Rovers had been supplied to all but two countries in the world, and military orders had been received from no fewer than 140 territories.

In the meantime, however, British military needs were changing. Equipment to be carried was becoming more and more complex, bulky and heavy, and even the 109in GS Land-Rover, with its ¾-ton capability, was sometimes inadequate. The need for extra carrying capacity led to the issue of another requirement – that for a new machine with all the agility and mobility of a Land-Rover, but with a minimum 1-ton payload. More than this, there was a subsidiary need for towing 4,000lb loads of various types, of which the new 105mm Light Gun was much the most important.

The demands of the Light Gun were too much for existing Land-Rovers (the towed/towing weight ratio was excessive, and weight transfer effects were also undesirable). The gun manufacturers were encouraged to consider driven wheels for the carriage, and up-engined Land-Rovers – 109in vehicles with the six-cylinder Rover 3-litre engine, and appropriate transmissions – were also tried. This latter scheme foundered, however, because the towing vehicle had to weigh less than 3,500lb, so that it could be airlifted by one of the latest powerful helicopters, and because the Gunners demanded more stowage space for their kit and stores. There was nothing for it except a new forward-control machine.

FVEE laid down the specifications – desired wheelbase, track, approach and departure angles, and payload capability – but then left the industry alone to make its own plans. It was a formidable challenge, though not one to defeat the people at Solihull. They knew that other makers would be invited to put up prototypes, and in particular they feared the opposition from Volvo with their 4140 series, but they had two advantages. Land-Rovers were now trusted friends of the British forces, and there was also a dossier of testing done at Solihull on powered-axle trailers in previous years.

A forward-control Land-Rover, with choice of petrol or diesel engines, was in production at the time (1970–1), but this could not meet the army's needs. It was too heavy, even when reworked

Sir Winston Churchill with his own 8oin model which he collected on his 8oth birthday

HM the Queen's Land-Rover special, in use at Stoneleigh Royal Show

Series IIs and IIAs had reshaped bodywork, notably along the sides. This IIA is being demonstrated to Lord Stokes at the Royal Show, Stoneleigh

to military standards, and by no means powerful enough to haul the big trailer loads across the rough terrain the army might encounter. There was no alternative, therefore, to a completely new 4 × 4 vehicle – as different from any existing Land-Rover as could be considered. Even the 109in and 110in civil forward-control vehicles retained at least 75 per cent standard Land-Rover components, but there would be no place for such standardisation in the new design.

In the terminology of racehorse bloodstock breeding, the new design was undoubtedly 'by Range Rover out of Land-Rover', as there is certainly quite a lot of each breed to be seen. Indeed, without the existence in production of the super-refined Range Rover (see Chapter 7), the army's latest forward-control device would not have evolved as it did. The kernel of the whole design was the availability of the Range Rover's very powerful power pack, 3½-litre vee-8 engine and transmission. It meant that 130bhp could be offered in a machine that need not be too heavy, as the big engine was a masterpiece of weight control. Once it became clear that this would do the job, the new '101in' vehicle came into being.

It appeared in public for the first time at the 1972 Commercial Vehicle Show, decked out in military colours. Its general nature, equipment and fittings, in fact, illustrated that it could not have been presented in any other guise. The cockpit, particularly, is quite unsuited to civil needs. Twenty years ago, when Land-Rovers were still stark, the 101in model might have sold commercially – but not today. Space is limited, even cramped, and equipment sparse and utilitarian. Seats are there to be sat upon and not to be admired. Even by Series III Land-Rover standards, the 101in model is not a thing of beauty, and by comparison with the Range Rover it is a sturdy but definitely ugly duckling.

Range Rover engines and transmissions are used, virtually without change, but there is no similarity in chassis design. The fact that the wheelbase lengths are merely 1in different is purely coincidence; the 101in was fixed by military requirements, and no Ranger Rover items are used. The chassis itself is enormously strong but simply laid out, and the engine finds itself almost between the seats. Transmission details, including the Range Rover's lockable centre differential, are unchanged, but there is no question of ride levelling being used at the rear. Nor is the Range Rover's coil spring suspension used – the military 101in relies on simple and sturdy half-elliptics.

At the time of design the powered-trailer concept was very important. It was thought that the 6 × 6 capability would be significantly better across very slippery terrain than that of a 4 × 4 towing a dead weight. And so it proved. Rover have released some remarkable film, which shows 101in machines towing Rubery Owen powered trailers through the most formidable-looking stretches of swamp, water and mud. With trailer drive disconnected the Land-Rover is defeated for lack of traction, but with trailer drive hooked up the grip seems limitless. In 6 × 6 form the hill-climbing performance borders on the miraculous.

Yet, for all that, trailer drive is no longer of the highest importance. Engineers and students of geometry will realise that the success of trailer drive is concentrated on the vehicle/trailer hook. In one confined position the towing, steering, articulating and driving function must all be centred, and if the design is not perfect, the consequences can be serious. In terms of extra performance for money spent, at this stage a trailer drive still has some way to go before it is proven and justified.

The first 'off-track' 101in military Land-Rovers were delivered at the beginning of 1975, and apart from minor sales to export countries, all so far built have gone to the British forces. The 101's baptism of fire came in the Trans-Sahara expedition of 1975, when the first four examples delivered were sent off to pioneer the journey from the Atlantic coast to the Red Sea. They came through with flying colours (see p 68).

Military Land-Rovers are also sold to foreign forces. Indeed, they have already been supplied to more than 140 governments. Nevertheless, if approaches are made from politically delicate quarters, reference is made to the Foreign Office and other interested bodies before the orders are filled. During the last Middle East conflict, for instance, an urgent request was received at Solihull for new supplies from one combatant – a request immediately sat on by British advisers in the interests of impartiality.

In view of Rover's position as a supplier of military transport, it is surprising that Range Rovers do not figure in army operations, since they would seem ideal cars for senior officers. The British forces, however, see no immediate niche for Range Rovers. Apart from their price – double that of the 109in Land-Rover in similar form – they feel they are too good for use across country, and rather too elegant for use as 'on the road' transport. How similar their conclusion sounds to that expressed by many Range Rover owners! At one point, certainly, the thought of combining a commanding officer's Land-Rover with his 'mufti' staff car looked attractive to the army, but after the thought of using a much-travelled and possibly battered Range Rover at ceremonial occasions had been considered, the idea was dropped.

The Range Rover's strong suits are performance, refinement and comfort. The forces find that their special Land-Rovers have enough performance and adequate refinement, and they are not paid to complain about lack of comfort! From time to time the planners look longingly at Range Rovers, but each time they turn away, thinking of economics and cost-effectiveness. The Range Rover is just too good for any job it could perform for the army, which is praise enough for any car.

Like the 4-ton Bedford trucks that are now a permanent part of the scene, the Land-Rover family is now an accepted part of the services' make-up. Although its technology dates from the 1940s, it has been much improved over the years, so that it still performs as well as ever and retains its unrivalled versatility. What is more, there is still nothing else on the market at the right price to match it!

The sales figures speak for themselves. By the end of September 1975, when Rover's financial year was completed, 962,414 Land-Rovers had been built. Of these, no fewer than 74,000 had gone to British 'government sales', which includes everything from military to Home Office use. A large proportion of these would have been campaigned across the world in khaki, navy blue or RAF blue, their lives sometimes exceeding a creditable 10 years of continuous service. No fewer than 42 per cent of all Land-Rovers sold in Britain since 1948 have been for government use. If anything like the same proportion applies to sales in other countries, it means that around 400,000 Land-Rovers have had some official or military use at one time or another.

Would the world have been a less well-ordered place without the machine? It would be nice to think so.

4

THE WORLD'S MOST VERSATILE VEHICLE

The plates on pages 97, 98 and 119 show Land-Rovers with a very wide variety of fittings, completely justifying Rover's claim that the Land-Rover is the 'world's most versatile vehicle'. It also illustrates Rover customers' opinion that a Land-Rover does not necessarily sell for what it is, but for what can be imposed on it. Even Maurice Wilks underestimated this potential, though he could hardly be expected to have known his little workhorse would have been converted into so many odd-looking vehicles.

It was the 'Go anywhere, Do anything' advertising of the early days that probably led to so many adaptations. People were tempted to test Rover's claims, and were not usually disappointed. The combination of four-wheel drive, a strong basic structure, and seeming indestructibility made outrageous demands feasible, and difficult ones a matter of routine.

Nowadays we expect any expedition to take Land-Rovers as a matter of course, and no Royal occasion to be complete without one or two; and we find nothing unusual in the thought of Land-Rovers being used on board British aircraft carriers. They have, after all, been everywhere else. Down a mine helping to dig for precious metals. Why not! Being waterproofed and driving across rivers with only the driver's head and a snorkel tube above water. Of course! Turning up in the Darien Gap where only animals had been seen before. Naturally! Carrying overhead inspection cages. What else! Putting out fires. What could be better!

Rover themselves rarely make such specials. They merely sit back and wait to be amazed by the latest improbable scheme that is proposed. The flow of ideas shows no sign of drying up. We have seen how the armed forces have used suggested models, but private firms are just as enterprising. Rover themselves went a long way towards oddity in the 1960s with their special machine for carrying the Rover-BRM gas-turbine racing car. Not only did this machine accept the racing car on a much-modified forward-control chassis, but it was confined to front-wheel drive only, and had a 'kneeling' rear suspension so that the turbine car could be driven aboard!

Perhaps the Land-Rover first made acquired worldwide fame in expeditions. Every owner soon found out that his machine could be used for just about everything, but it took long-distance treks like the early Oxford and Cambridge expeditions to bring in the publicity. Whether it was Trans-Africa, London to Bombay, or trips up the Amazon with students aboard, there was usually a Solihull-prepared Land-Rover in the party.

After a time this business became so overdone that Rover found it embarrassing; they discontinued the preparation of specials, and decided to confine their help to advice only. Not that this was given on a sketchy basis. Over the years they had built up so much know-how on the art and craft of crossing unknown territory that they were able to publish a useful little book called *A Guide to Land-Rover Expeditions*. This book, which is updated from time to time, contains the widest variety of information. For instance, on driving methods it may state, 'Before fording make sure the clutch housing drain plug is in position and, if the water is deep, slacken off the fan belt', and on medical care, 'The danger of snake bite tends to be exaggerated'. Remarks on food include, 'There is no need to be unduly squeamish about eating strange dishes', and advice on laundering contains the following paragraph: 'A convenient way of washing clothes whilst travelling is to put them in a waterproof, sealed, container in the back of the vehicle with a suitable amount of water and washing powder. After 100 miles they should be clean! Water for rinsing must of course then be available.'

This little book, in fact, illustrates what hazards a Land-Rover may have to face on expeditions. Rover advise, even if this *is* a Land-Rover, it should still not be overloaded. They also give a long list of desirable extras, a breakdown of the tyres most suitable for different terrain, a recommendation for extra tools, and advice on spares and maintenance. Tips on driving in 'off the road' conditions include such items as,

> Never wrap your thumbs round the steering wheel rim. If the vehicle hits an obstacle the steering wheel could be jerked so hard that the spokes could catch and break your thumbs. Many people have learnt this the hard way.

There is advice on currency strategy, diplomatic detail, immunisation procedures – everything, in fact, that a novice traveller might not know. Right at the end, though, is the following counsel:

> The Land-Rover has been associated with expeditions for many years, and the publicity we can gain from these and similar ventures is now very minimal. Consequently, as a rule any request for monetary or material assistance will not be considered.
> As a guide, when approaching a potential sponsor ask yourself if you are offering him value for money. There are few philanthropists in the world.

Not that this stopped explorers from using their Land-Rovers. Brockbank once produced a cartoon showing a team of mountaineers struggling to the top of a peak, only to find a Land-Rover sitting there waiting for them. Rover themselves, of course, used the vehicle's abilities to support such advertisements as the 'No through road for motor vehicles' sign, above a new sign that stated proudly, 'Except for Land-Rovers'.

Land-Rovers climbed Ben Nevis, which was good for publicity

at the time, until someone pointed out that a Model T Ford fitted with half-tracks had done the same thing in 1911. 'Maybe so', was the retort, 'but the Land-Rover was standard, and didn't need man-power to help.'

Even in temperate Britain, there are parts of the country where any winter journey might turn into an expedition. Is it any wonder that post office, AA and RAC transport, and emergency and other essential services all employ Land-Rovers? Go to the far north-west of Scotland, and you will see little else. Go into the wilds of Wales, where a postman spends most of his time off the public highway, and the bright-red delivery service takes on a Land-Rover shape. If this sort of thing is normal in Britain, you can be sure it is normal in rougher terrains. Look at any movie of an African safari, and you will find that the rhino always charges a Land-Rover. The Land-Rover survives every time, and the rhino usually has a headache. Land-Rovers may not have replaced the St Bernard dogs in the Swiss Alps, but their day will come!

In recent years four-wheel drive Rovers have taken part in all manner of long-distance trips. The bridging of the Darien Gap, between Central and South America, brought fame to Major John Blashford-Snell's army party, their Range Rovers and the Land-Rover that was flown in to help. The expedition received much publicity in 1972, but Richard Bevis and Terry Whitfield had covered the same ground in 1960 – in a Land-Rover!

The earlier expedition used an 88in Series II vehicle, fitted with the van tail, and started out from Toronto in Canada. It set out from Panama City in February 1960, and it took the two men and their Land-Rover no less than 134 days to hack, drive and winch their way through the jungle into Colombia. They had no generous sponsors, no radio links and no airborne support, so they can certainly be said to have done it on their own. Richard Bevis stated that they built 125 palm log bridges during their journey through the jungle, and they suffered no fewer than ninety punctures!

When Major Blashford-Snell's assault group set out on an Alaska-to-Cape journey in 1971–2, using Range Rovers, the whole project was on a more heroic scale. In theory the Alaska–Panama, and Colombia–Cape Horn sectors should have been a very pleasant drive, with all the dramatics taking place in that ferocious jungle east of Panama City, but there was trouble in Canada when one of the Range Rovers hit a lorry on an icy road. Conditions in the jungle were as bad, and worse, than expected, and even with experienced and determined military members of the team, progress was slow. For all the excellence of the Range Rovers, it is known that they are not quite as all-conquering as Land-Rovers in impossible conditions, and there were some unfortunate breakdowns owing to overloading with extras and equipment.

Blashford-Snell eventually found a battered old Land-Rover in Panama, had it flown in to a tiny airstrip in the middle of the jungle, and used it as a pathfinder thereafter. Not that this is any criticism of the Range Rovers, which were much bulkier and much more ponderous to manhandle round obstacles. The Land-Rover was invaluable, however. In his book, *The Hundred Days of Darien*, Russell Braddon has this to say of the vehicle after it had capsized into a deep ditch:

> Setting up their Tirfor jacks, attaching ropes to the apparently lifeless Land-Rover, they winched it upright and hauled it diagonally back on to the track. Never a thing of beauty, it now looked more derelict than ever. But the Land-Rover is tough and without conceit. As if capsizing into jungle ditches were something routine, she responded to the first touch on her starter.

By the time the party had crossed the Isthmus of Darien in April 1972, there were few unconquered parts of the world left for a Land-Rover to tackle. Therefore, when another military team were looking for ways to prove the worth of their new 101in vee-8-engined Land-Rovers, they faced a problem. In addition to Darien, trans-continental car rallies had shown that there was no longer any challenge in Asia or in South America. Their decision was to tackle the Sahara desert – not from north to south, which is now a thoroughly hackneyed trip, but all the way across, from west to east. As far as the Ministry of Defence knew, this crossing had never before been achieved, and a 7,500-mile route from Dakar on the Atlantic to Cairo on the Suez Canal was chosen.

Squadron Leader Tom Sheppard of the RAF led the expedition, whose object was not only to make the first crossing, but also to prove the worth of Solihull's latest and rather remarkable machine. Tom Sheppard, in his *Autocar* story of the successful expedition, had this to say about the Land-Rovers:

> The 109in long-wheelbase Land-Rover's limitations in extreme conditions are easily found. Thrashing over deep rutted tracks most used by vehicles two or three times its size, its underbelly clearance is frequently the penalty paid for its light agility ... But Rover's new 101in wheelbase 1-ton military Land-Rover is a unique combination of optima that, market-wise, sits precisely in the middle of the seesaw of payload and power/weight ratio.

So it proved. The services took along four of the new machines, two of them equipped with Rubery Owen's powered trailer. Much of the terrain was known in theory, but in the middle came the vast Mauritania/Mali 'Empty Quarter', where the available maps were poor and the terrain worse. Mile after enervating mile was just pure sand, deep and clinging, where aluminium sand-ladders were essential, and the additional grip afforded by the powered trailers invaluable. Petrol supplies were nearly non-existent, and at one stage the team had to travel nearly 1,200 miles between petrol points. In such conditions the petrol consumption was daunting, anyway.

With the powered trailers giving six-wheel drive, the Mauritanian sand at its dragging worst, and very heavy loads being carried, Rover themselves had estimated for less than 8mpg; and in the worst conditions consumption rocketed to 5·7mpg.

Compared with the Darien assault, there were few dramas, and there was also no need to add human muscle and outside help to the performance of the sturdy machines. The latest 101in Land-Rover uses numerous Range Rover components – engine, trans-missions and axles – but there were no mechanical breakdowns; the problems encountered on the Darien expedition – particu-larly with regard to correct vehicle loading and correct lubrica-tion – had quickly been learned and overcome. The entire Trans-Sahara trip took just 100 days – an average of 75 miles a day – and it meant that a Land-Rover had eliminated yet another of the world's 'impossible journeys'.

Perhaps, after this, there really are no other natural barriers worth attacking. Siberia perhaps, if the Russians would allow it, though Prince Scipio Borghese and his Itala cracked this one as long ago as 1907, in winning the Peking-to-Paris race. The North Pole or the South Pole? In the 1940s, perhaps, but not in the 1970s, when even the Poles have been extensively colonised. Perhaps the Land-Rover, and its fellow 4 × 4s, have tamed the world.

People who have tackled the 'impossible journeys' have usually done so in near-standard machinery. The Land-Rovers used certainly carried many extras, particularly the essential winches by which a stranded vehicle could get itself out of trouble, but they were still recognisable as Solihull products. The same could not always be said about some of the strange vehicles built up on the Land-Rover base; this sort of rebuilding has not yet been practised on Range Rovers, perhaps because of their price, but the potential is there.

The logic behind the many complete rebuilding jobs carried out on Land-Rovers is simple. Rover's claim that the Land-Rover is the 'world's most versatile vehicle' tells all. No builder with a special vehicle in mind need look any further. Over the years, Rover's publicity staff have issued photographs of Land-Rovers converted into troop carriers, armoured cars, or mobile ambulances, and they look positively mundane by now. The Land-Rover decked out by the police for patrol, breakdown and rescue work might be brightly painted in its red or orange striping, but is still nearly normal.

Tom Barton has said that the strangest conversion he ever saw was the original central-steering prototype, but admitted he was often surprised by the things his machines were asked to do. My own favourite was the new Series IIA machine converted to running on rails, and depicted pulling trucks loaded with a number of other Land-Rovers. Even if 70bhp and four-wheel drive is not nearly enough to make a 'shunter Land-Rover' practical, the idea was interesting, and BR were intrigued.

It was run very close for novelty by the hovercraft Land-Rover that Soil Fertility produced, which was designed to allow the Series II's weight to be reduced when it had to spray fertiliser or weedkiller over a seed bed. In this case Solihull was happy to release a picture showing a normal Land-Rover bogged down, and the 'Hover-Rover' floating serenely past it.

Road sweepers and snow-plough attachments look entirely appropriate, and even miniature fire tenders sit suitably on a Land-Rover chassis. The larger forward-control chasses are best for these jobs, as their maximum payload is so high.

I have seen Land-Rovers carrying air-stairs for plane-to-tarmac airfield use, and I have seem them equipped with baggage-loading carrier belts and adjustable elevating struts. I have seen both Land-Rovers and Range Rovers with twin rear axles (they looked so 'right' that a second glance was needed to work out what was actually different), and I have often seen coach-built versions that allow dignitaries to stand, or sit, in elevated positions. In 1975 Rover engineers built a 'special' Range Rover with removable roof panel and gun-aiming platform for a certain Middle East potentate – so that he could go shooting in comfort.

Another sheikh ordered a brace of specials, and when asked about the colour, produced a rather exotic flower petal, saying 'Copy that'. The problem was not in duplicating the colour but in ensuring that the petal survived the journey home to Solihull.

Local authorities have often invented new jobs for their Land-Rovers. There have been tiny little dustbin lorries with tipping gear and all the usual fittings, and elevating hydraulic platforms for street lamp inspection. A 109in Land-Rover with truck cab and open rear can be loaded with all manner of equipment: for instance, I have seen them carrying air-compressors, to service the pneumatic drills of a particular repair job. Donald Campbell used Land-Rovers to provide start-up power for his turbine-powered Proteus-Bluebird in the 1960s, and when the Shell Oil company were developing fuel cells as an alternative to petrol engines, they mounted the cells in a Land-Rover and drove the axles via electric motors. This experiment proved that fuel cells were so bulky that there was no payload left to provide a useful carrying capacity.

Chopping and shortening chassis frames has often been a popular business (the rear propeller shaft is then very short), so that the Land-Rover can act as a tractor for various trailers. Horse-boxes, trailer workshops, mobile radiography units, and even mobile homes have benefited from this conversion, though caravans are more usually built on an unmodified chassis.

Land-Rovers can provide the base for mobile cinema screens, complete with removable loudspeakers, of course; and for golf ball development, using the power take-off to run a sling. They can also carry any sort of equipment wherever normal cars cannot penetrate; the Road Research Laboratories' soil-sampling rig, on the back of a 110in Land-Rover, was one good example. In addition, Land-Rovers with enormously wide axles, very fat wheels, and the air of a young boy putting on weight far too fast, have been seen on Forestry Commission land.

For the most enterprising conversion, the award must go to the tracked conversion produced by Cuthbertson. For military use, it was based on a 109in Series IIA normal-control Land-Rover, and had tank-type tracks driven from each of the four wheels. Ground clearance must have been at least 15in, and within the limits of gravity and engine power this device could go anywhere. The British Army wanted it for bomb-disposal work, though it is not true that they did this merely by driving the indestructible Land-Rover over the bomb itself. Not that bombs could destroy the vehicle anyway. Only recently TV news film from Portugal showed a Land-Rover that was said to have been blown 6oft through the air. The bodywork was riddled with holes and the tyres destroyed, but apart from this it looked as if it could have been driven away for repairs.

5

REFINEMENT AND COMPETITION

In the beginning, you could never accuse a Land-Rover of being civilised. The company would certainly have been insulted, and the car – well, that would have been damning it with faint praise. Land-Rovers were not meant to be *nice*, after all – they were designed to be effective, and for years customers bought them purely for what they could be made to achieve. Comfort, silence, warmth and high performance were not Land-Rover attributes, for they needed to make no concessions. But then they started to become fashionable.

Owners started using them to pull horse-boxes to point-to-point meetings, to travel to and from a day's shooting, and as mobile grandstands at all manner of functions. Some even had 'rough and smooth' Land-Rovers, one as a working machine and the other as an 'estate' car. Once people started converting them into caravans, and drove long distances in them, they began to demand more comfort.

The sales staff at Solihull were not too happy at this development, for they had seen how other 4 × 4s had lost their edge as working machines. If the Land-Rover was to become civilised, they were determined it would lose nothing else in the process. In the end, of course, Rover brought out the Range Rover (see Chapter 7), which is what the Road Rover might have been in the 1950s.

It was not British competition that forced changes, for nobody, not even the established tractor manufacturers, tackled Rover at first. Austin's Champ was purely a military machine, as we have said, but even though it was dropped by the Army in favour of nearly standard Land-Rovers, Austin decided to try again. If only their Gipsy had been more versatile, and sold in higher quantities than it did, Land-Rover development might have had to be speeded up.

Austin's Gipsy arrived on the scene early in 1958. It was sheer bad luck for Austin that a vast range of much-improved Series II Land-Rovers came along just two months later! The Gipsy followed normal 4 × 4 design practice in many ways, with rear-wheel or four-wheel drive to choice, a low range of gearing for exceptional jobs, and such options as rear power take-offs. It differed from the Land-Rover in that its bodywork was made of pressed steel, and it had all-independent suspension.

Wheels that moved up and down independently were all well and good in theory, but Land-Rover designers knew that they could lead to trouble on really uneven ground. When a Land-Rover's wheels rose on full bump, the axle casings rose with them, but with independent springing the cases stayed where they were, vulnerable to damage from centre ridges and rocky outcrops. As for the steel bodywork, it was subject to corrosion. Rover's own choice of light alloy had been Hobson's Choice in 1947, but was now a sales advantage, for light alloy does not corrode, and where but in agricultural and rough use was corrosion more likely?

That apart, the Gipsy's problem was that there was only one basic model and a choice of engines at first – no station wagon, not even a hardtop option. It was similar in size to the Land-Rover but always more expensive. Later a long-wheelbase option appeared, and in 1962 independent suspension for the new range, which had beam axles and leaf springs, though one version kept its independent springing at the front only.

Rover were still unworried by this domestic competition, but had to take note of its trim changes. Not only were the instruments in the correct functional place, ahead of the driver, but the

The forward-control Land-Rover at work. It was in such application that the 25cwt cross-country payload was so useful

The special 'lightweight $\frac{1}{2}$-ton' Land-Rover made for the military. This was one of the first vehicles, built in 1968, with headlamps fitted into the nose. On the lightweight model, the doors, screen, seats and top could all be removed and parachuted as a separate load

The 'lightweight $\frac{1}{2}$-ton', proving that it was air-transportable in stripped-down form, below a Wessex helicopter

facia looked as if it had been drawn by a stylist. There was a glove locker, trim pads on the doors, map pockets and arm rests. In the end such refinements did not save the Gipsy. Production came to an end at the beginning of 1968. With much superior facilities, and eventually even with a wide range of models, Longbridge had not been able to match the record, nor the mystique, of the Land-Rover.

In the meantime Tom Barton's engineers had been very busy; the idea of a 'luxury Land-Rover' had finally become respectable. Improving the trim and furnishings was quite easy, but what about the mechanical specification? The whole question revolved around engines. The 'fours' were robust and seemingly ever-lasting, but if the Land-Rover was to be refined, they would have to go. The equation 'refinement = smoothness' would have to be satisfied, and this meant using more cylinders.

In the early 1960s, when the project began to take shape, Tom Barton was faced with Hobson's choice all over again. Apart from his existing power units, the only Rover-built alternative was one version of the straight 'six' fitted to Rover's passenger cars. This was directly related to the original Land-Rover petrol engines fitted up to 1958, having the same unique valve arrangement and construction. As a design it was already old, for Jack Swaine had started designing it before World War II, and the first had been sold in 1948. Two sizes were being made – the 2·6-litre and the long-stroke 3-litre.

Quite suddenly, in 1964, there might have been another choice. William Martin-Hurst pulled off his spectacular deal with General Motors for Rover to take over production of the light-alloy vee-8 engine from Buick. This engine would not have been suitable for the Land-Rover – with 3½ litres, and any choice of power output above 130bhp, that was taking things a bit too far – but what intrigued Barton was that something might spawn from it.

Jack Swaine and his engine designers had similar ideas. Since the 1930s Maurice Wilks had carried a liking for the vee-6 layout in his head, and over the years a number of experimental engines had been built. All had been shelved, but with the sudden availability of new hardware the old concept came back to life. Maurice Wilks was dead, but his ideas lived on. Jack Swaine told me,

> This was really because we knew that Buick had had a cast-iron vee-6 and incidentally we had great difficulty in persuading GM's Joe Turley that we had made experimental vee-6s in 1951! They were equally convinced that Buick had invented a 90-degree vee-6. We thought that the same approach would be very suitable for Land-Rovers, and in fact we did a design based on the bare bones of the Buick vee-8 engine. It would have been ideal for its purpose, and it would easily have fitted in the engine bay, but the extra cost of tooling killed it as a project.

Barton, therefore, was left with the existing six-cylinder engine, and set about squeezing it into the Land-Rover's structure. This was surprisingly easy, confirming once again that much of the bulk in modern engines is due to accessories and 'hang-on' items, for a look under the bonnet of the vehicle shows little obvious sign of crush. A bit of pushing and shoving here and there ('We were very clever in reducing the clutch and bell-housing space') made all the difference, and without major expense the company suddenly had its alternative engine.

This even made sense to the production planners, because when the new Rover 2000 had ousted the old P4s, and the 3-litre had been re-engined with a vee-8, there would still be an outlet for the old unit. It was necessary to make allowance for new details, the 'softened' engine tune, and a reduced demand, for the new line to be ready. The management was delighted.

So, too, were the customers. They had first been offered the new silky 'six' in the uprated forward-control Land-Rover at the Commercial Motor Show in 1966, but its fitment in the normal (long-wheelbase only) machine followed in April 1967. This completed a very impressive range of petrol and diesel-engined machines, with all manner of bodies, ranging from the simple

78

pick-up to the twelve-seater station wagon. As ever, the Land-Rover had to suffer the quirks of the British purchase-tax system. A basic machine was free of tax, but a 109in ten-seater station wagon was classed as a private car and taxed – unless an extra couple of seats were squeezed in to make it a twelve-seater – a bus in the eyes of the law – when it was free of tax again!

The optional six-cylinder engine was a considerable bargain at an extra £60, and as soon as it appeared, the facia was tidied up, and given the luxuries of a key starter arrangement and linked wipers. There was still no evidence of fancy trim, however, and every screw, bolt and fastening remained clearly visible. In July 1967 the company offered an optional 'luxury front seating' pack. The seats, marketed only as a matched trio, cost £53, and made a great difference to the looks, if not to the actual comfort, of the driving compartment.

The next, and most noticeable, change to the Land-Rover was forced upon the company by legislation. Ever since the vehicle had been announced, it had carried headlamps closely fitted to each side of the radiator grille – at first, even, covered by protective mesh. From the spring of 1968, however, the headlamps were moved outboard, taking up a new position in the front of the wings, where they still are. This was to satisfy certain new legal requirements in overseas territories regarding the closeness of headlamps to the outside of the car. Even without the stimulus from abroad, new British law was also in the wind, so a change would have been necessary, even if only for the domestic market.

Finally, in October 1971, Series IIA vehicles at last gave way to the latest Series III machines. The change came when total Land-Rover sales were approaching 800,000, and after a record sales year with the last of the Series IIAs. There was certainly no need to make changes to revive a flagging model. In that financial year (1970–1) no fewer than 56,663 Land-Rovers of all types – most rolling off the lines ready to work, but some being shipped overseas, crated and CKD, for local assembly – had been made, still a record at Solihull. Not only were there many thousands of 'standard' models, but hundreds of lightweight '$\frac{1}{2}$-ton' Land-Rovers for the British armed services and rather fewer of the big 30cwt forward-control devices, which were about as different from a normal machine as one could get.

The Land-Rover had then been on sale for more than 23 years, easily a record by any previous Rover standards and demand simply continued to rise. Workmen and management could not remember a time when the lines were other than flat out, when a production shortage of parts was not more desperate than a shortage of orders, or when some boost was actually needed. Since 1961 nearly half a million of the Series IIA machines had been sold, and this to a market where the original Land-Rover purchases resolutely refused to wear out. It was one thing selling masses of cars to a market whose old models would have rusted themselves into a scrapyard, and quite another when the old models continued to work for ever and a day. To Rover the phrase 'planned obsolescence' was something only practised by their competitors; they knew very little about it.

It was not easy to see any difference between the new Series III Land-Rover and its predecessor, for the only exterior changes were a new grille and a new badge; but it was much improved, both mechanically and in its cab. For the very first time the Styling Department had found time for a look around the vehicle. As Tony Poole of the department said: 'We looked, and we looked, but did very little. As a design for its own job, the Land-Rover was so "right", so suitable, that it was nearly impossible to make simple improvements. With a lot of money for new tooling . . . sure, *anyone* could make improvements on that basis. But on the Land-Rover it simply wasn't necessary.'

In the cab of the Series III there were no changes down below, so to speak, where there were the usual profusion of gear levers, handbrake sprouting from the usual place, and pedals big enough to handle the largest foot; but above that there were vast improvements. There was a new steering wheel, for a start, with a centre boss that looked as if it housed a horn push, but did not; and in front of it, looking very smart and quite out of character, was a very neat new facia. In addition, padded facia rolls, headlamp flashers, proper screen ventilation and a fresh-air heater were standard. The de Luxe seats were optional, as before, but it seemed that all the protruding bolt heads, the sharp edges, and the icy draughts of yesterday had been eliminated. There was even provision for a radio in the new facia, which gave a clue to the sort of custom a Land-Rover was now attracting.

Under the skin, the changes were more subtle, but just as important. The basic permutations on wheelbase lengths, engines, body styles and standard extras remained, but the new vehicle's gearbox was 'all-synchromesh' and there was also a diaphragm clutch. Brakes were bigger and stronger, and some long-wheelbase machines had brake servo assistance. A stronger rear axle was fitted, and more efficient alternators replaced the old type of dynamos.

The motoring magazines rarely got their hands on a Land-Rover, so any road test was bound to be of interest. *Autocar* tested a 109in six-cylinder Series III vehicle, and had several interesting points to make:

> There were few adverse comments on the cab interior, which proved draught-proof and was trimmed to a standard unheard of in Land-Rovers a few years ago ... It is amazing where the Land-Rover will go, even in two-wheel drive and high gear ... As long as the ground is dry there seems no limit to what the Land-Rover will climb in low ratio first gear. It will certainly go beyond what the average driver's nerve will stand. It can also be taken *across* a steep slope with impunity, tilting a long way from the horizontal with no risk of falling over (Rover claim a full 45 deg; we didn't substantiate it) ... The feeling that it will go where nobody else can follow is curiously comforting, even if one can rarely take advantage of it.

Since then only one significant mechanical change has been made, and that was quite unexpected. Rover at last agreed with some of their customers about a Land-Rover's low gearing, and offered an overdrive as an option. Overdrives for Land-Rovers cannot be the same sophisticated and smoothly operating units found in private cars, for, owing to the mechanical congestion at the rear of the main gearbox, there is just not enough space. Tom Barton told me that when the option of an overdrive began to look desirable, he invited several of Rover's approved accessory suppliers to put up schemes. It was Fairey, who make winches already, whose design was chosen, and is now optional.

The Fairey overdrive is a simple little two-speed gearbox, which is set directly to the rear of the main transfer box, uses Rover 2200 synchromesh components, and can be fitted to any post 1958 Land-Rover in about 3 hours. The only 'blacksmith engineering' necessary is that a hole has to be cut in the floor to accept the gear lever. When in use the general gearing is raised by nearly 28 per cent, a really large upgrading, and operates on all gears, including reverse. Unlike the conventional Laycock overdrive, where cone clutches and epicyclic gearing emphatically do *not* like going backwards, the Fairey Winches item is happy to do its job in all conditions. Theoretically, of course, your overdrive-equipped Land-Rover has a choice of sixteen forward gears and four reverse ratios, if all the combinations of high and low ranges, direct and overdrive transmission are considered, but in practice the overdrive is only used in High Range.

When I first stepped into the cockpit of an overdrive-equipped Land-Rover, I was rather confused. Instead of the usual floor-mounted array of three brightly painted gear levers, there was an extra one, and for anyone but an expert Land-Rover driver, it was difficult knowing where to start. I found it easiest to choose the normal High Range, or the 'overdrive' High Range, before moving off, selecting the appropriate settings and firmly ignoring that extra lever. It would be possible to swop ranges while on the move, rather like a well-drilled juggernaut driver playing with all his ratios, but I did not think that was what Fairey had in mind. The edge of performance was lost, of course, with the overdrive in play, but fuel consumption seemed to be improved by about 3mpg.

There, for the moment, the specification seems to have settled down. Persistent rumours of a 3½-litre vee-8 engine option can probably be discounted, as Range Rover transmissions and axles would also be needed to withstand the vastly increased engine torque. Specials have indeed been built, and they are used in hill rallies and other forms of 4 × 4 sport. Such machines may be in evidence at Solihull, but even if the planners wanted to make and sell some, they would have the usual problem of finding enough space to produce them.

The Land-Rover is as much in demand as ever it was, but Rover must now be wondering about the 1980s. British Leyland now find themselves in the classic 'long-running-success-story' dilemma. Should they now be spending time and effort in preparing a radically new Land-Rover, when their sales staff are sure they do not need one; or, if signs of decline are not apparent well in advance, will they be caught unprepared when a new model is needed?

We have, of course, all been asking these sort of questions since the 1950s, and they are perhaps more applicable to a private car than a working machine. The Land-Rover's engineering needs no modernising, though a casual look at the modern Land-Rover might lead one to assume that much of the 1948 design survives. But this is not so, as Tom Barton has explained:

Nothing from the first Land-Rover, not a single component or casting, is still used in the latest Series III machines. Nothing, that is, except for one or two standard nuts and bolts. Everything else – chassis, engine, transmission and bodywork – has changed completely over the years. It has all happened very gradually, as you can see from the records. One year we might alter chassis and wheelbase lengths, another year there would have been a major engine change or new engine option, and at another time we would have gone in for a minor re-style.

The Land-Rover is in a similar situation to the Volkswagen 'Beetle'. The current Beetle and the 1946 Beetle might look the same, but the details are now completely different. Rover, like

83

Volkswagen, believe that the concept of their machine was right, but in detail they have always been looking for improvements. When something came along which was good for the customer, good for the company, and an all-round improvement, it would be specified.

I once asked Tom Barton whether anything fundamentally wrong had been committed to production, and after some thought he said:

> On reflection, perhaps, right at the start, we didn't realise just what the public would ask their Land-Rovers to tackle. We under-estimated their faith in the vehicle. We didn't know that they would expect their Land-Rovers to do *anything*! That explains why we have done a lot of work on engines and transmissions over the years. At first they were just very strong – now they are nearly bomb-proof. In other ways we have made a strong structure even stronger, for the same reasons.

The fact that Rover is now part of the post-Ryder British Leyland set-up has made planning even more difficult to forecast than usual. Production experts rarely have time to look beyond the next demand for increased production, and right now the sales experts do not need to. Styling chiefs like David Bache and Tony Poole also usually plead lack of time; and designers like Tom Barton and his men usually have a sheaf of sketches and brochures of ideas tucked away, and say nothing. Gordon Bashford, who has now moved on to higher things in Spen King's 'behind closed doors' department, said: 'If I was starting again with a new package, I would start from the Range Rover chassis with coil springs, and I might want to style a new body on the same lines. A vee-8 version would be interesting too.'

No matter what the Land-Rover of the 1980s may look like, it will undoubtedly continue to make Solihull burst at the seams. Even though the vast new assembly hall at Solihull will open up new possibilities – P6 assembly will eventually move into it, and leave the North Block available for new Land-Rover and Range

The 101in forward-control army Land-Rover, powered by the Range Rover's vee-8 engine, here seen towing the Rubery Owen powered trailer. Effectively, this arrangement gives a six-wheel drive, articulated vehicle, which can cross anything as long as the wheels can grip

When the army had second thoughts and asked for the headlamps to be separately mounted, this was Solihull's way of checking the alternatives

(*Right*) You could take a Land-Rover *anywhere*, according to Brockbank

(*Below*) Land-Rovers were advertised in this way for some time

(Left) One of the 'Darien' Range Rovers inching its way down a 'swamp ladder' to cross a deep ditch

(Below) A 'Darien' Range Rover falls into a hole in the riverbed while fording a fast-flowing torrent

Rover assembly facilities – there will still be no let-up for the planners.

With such a reputation as the Land-Rover's in almost every country in the world, I cannot see any sort of freewheeling at Solihull. I said almost every country, for when I asked a salesman to list the countries in which Land-Rovers were sold, he said it would be easier for him to list those countries that had not taken any. They were Albania and North Vietnam, but Rover would not be at all surprised to receive orders for spares from these countries.

6

THE PRODUCTION JIGSAW

Do not show a production expert around the Land-Rover production lines and expect him to be happy. He might be amazed, he might be a touch incredulous – but he will not be impressed. Over the years Land-Rover production grew haphazardly, and every manager at Rover treats the Land-Rover buildings with a mixture of amusement and affection, and even at times with a spot of ridicule. By the standards of the 1970s the way in which vehicles are built is outdated. The whole of Solihull is not like this. Not far away is the Rover P6 model assembly block, brand new in 1963, while just around the corner is the vast, newly completed building where Rover's sleek new 1976 models are being made. In assembly methods, as in so many other ways, the Land-Rover is a law unto itself.

If you were to stand at the gates of the Solihull factory, you would be astounded by the intensive movement of loaded British Leyland trucks, many of which are servicing the assembly of Land-Rovers. In these times, when 'integration' is the last word in financial planning, Land-Rovers and Range Rovers are in fact built up piece-meal all over the district.

Naturally, on the assembly lines bodies meet chassis frames in the right places, and engines arrive punctually on their station by overhead conveyor, with pallet after pallet of springs, dampers, headlamps, nuts and bolts, glass, brakes, seats and details alongside. But engines, transmissions, axles, bodywork, and all the special bits and pieces come from elsewhere.

As we have said, when Spencer Wilks authorised the Land-Rover's introduction in 1948, he expected to sell about 100 of them every week. Nowadays the company could sell twenty times that number, if only they had been able to expand the facilities to build the vehicle at the same rate as demand. Today, at Solihull, there is very little space to spare and no impressive evidence of automation; a large part of the Land-Rover is manufactured elsewhere.

When Rover moved all their car-making operations into Solihull, from Coventry, at the end of 1945, they must have thought there was plenty of space for future expansion. The directors had even made sure of this by buying up all the surrounding farmland during the war years. But as Land-Rover sales boomed, 'Auntie' Rovers proliferated, the even statelier 3-litre saloon came into production, and the new Rover 2000 waited in the wings, the directors had to think again.

Every time Land-Rover production was raised (and Rover car sales boomed during the 1950s, too), more hardware had to be squeezed out of existing buildings. In such conditions more vehicles could only be produced by speeding up the process. There comes a time, however, when this solution becomes counter-productive, as well as irksome to the people who have to work it. The alternative is to add duplicate tracks, duplicate conveyors, duplicate storage areas – and to hire more men.

Without extension of the Solihull factory buildings, vetoed by Government policy for a good many years, it was clear that in due course increased production would eventually be halted by brick walls. Dropping the P4 'Auntie' Rover in 1964 helped a bit, for it had always been assembled under the same roof, but even before then desperate remedies had been needed.

The net result, after 20 years of forced diversification, is that Birmingham is liberally endowed with Rover 'satellite' factories, while some Rover-made parts come from as far away as Coventry. In 1975 there were eight Rover factories in the Birmingham area. In Coventry a small subsidiary factory at Clay Lane (just behind the site of the original Triumph car plant) made Land-Rover body parts, and even Alvis in Holyhead Road made Range Rover vee-8 engine components. Bits and pieces for the four-wheel drive machines flowed into Solihull, the largest factory of them all, from every single one of these.

The eight Rover factories in Birmingham comprise two from the original 'shadow production' scheme, an old Great War munitions works, and five smaller buildings bought since 1952. Each was an essential buy at the time, because the company was beginning to burst out of existing premises it already owned.

Rover's roots were in Coventry, where they once built cycles, motorcycles and cars under the same roof. After World War I they began to look around for another factory. In 1919 Jack Sangster persuaded them to build his own air-cooled 'Eight' design, found the redundant Munitions Components Ltd building at Tyseley, and brought them all together. Eventually this became 'Tyseley 1' in Rover language, the firm's first factory in Birmingham.

Then came British rearmament in the 1930s. Spencer Wilks's company was invited to join the motor industry 'shadow' factory scheme, and was allocated a brand new factory at Acocks Green – to build aero-engine parts. That was in 1936, and Acocks Green was only 2 miles from Tyseley. Three years later, with war imminent, Rover were allocated a second, even larger, 'shadow' factory that was being built on virgin farmland at Solihull, 3 miles further out of the city centre. These factories made up a logical compact group, and the Coventry factory began to look like the 'odd man out'.

Shadow factories were originally operated on behalf of the Air Ministry, but Spencer Wilks reasoned that the Ministry would have no use for them after the war, and that Rover might then buy them. It was on that basis that the management bought up hundreds of acres surrounding the Solihull plant. Without this purchase, modern buildings like the P6 North Block, and the big assembly building completed in 1975, could never have been planned.

Solihull, Acocks Green and Tyseley were big, but by 1950 they were not big enough. To secure space for expansion, the buying process had to begin again. First purchase was a factory in Perry Barr (1952) followed by another in Percy Road (1954), and this sufficed for the next 10 years. Once the success of the all-new Rover 2000 became evident, it set up a real 'house-of-cards' planning process. If Rover 2000 axles had to be built here, Land-Rover parts would have to be moved to there; if engines were to be machined here, other aspects of Land-Rover production would have to be shifted to there.

Another factory was erected at Cardiff, which helped, but it was not enough. In the 1960s, with further expansion and with the Range Rover project in mind, Tyburn Road (1965, bought from AEI), Garrison Street (1965, from Wrights Ropes) and Tyseley 2 (1969, from CWS Ltd, and conveniently alongside Tyseley 1) all were added to the roll-call of factories. These, among others in the West Midlands, are part of the huge collection of real estate that British Leyland have to manage. The financial press has suggested that British Leyland should close down their oldest and smallest buildings, but not mentioned the anti-expansion Government policies that forced them to buy these factories in the first place. Financial logic, with no thought for the work force, suggests that all operations should be concentrated on the three biggest factories – Solihull, Acocks Green and Tyseley (1 and 2). But a compassionate firm, and British Leyland is as thoughtful as most, would not even consider such a move.

With up to 1,300 Land-Rovers and more than 250 Range Rovers rolling off its assembly lines every week, Solihull's appetite for component parts is enormous. The only items made from scratch there are body pressings and the body shells that use them. Everything else – engines, transmissions, chassis frames, electrics, etc – arrives in truck loads through the gates. Ice, snow, fog, breakdowns, accidents, and strikes can all bring the lines to a halt in a few days, as storage space for deliveries is limited.

93

At Solihull it is necessary to define what an assembly line is. To most of us it is the line where power train meets chassis, where chassis meets body, and from which the machine is driven off the end. But Tom Barton said there was more to it. He mentioned build-up and trimming of bodies, and the fitting of such items as winches and power take-offs. That work was not done on an assembly line. The main assembly lines were just one part, an important part, of the process.

I took his point, but asked him how many of these were there. At Solihull there are rigid boundaries between certain activities. Land-Rovers for general use are built on three parallel main lines. In an ideally planned layout each line would concentrate either on 88in or 109in machines, but from time to time one line accepts a mixture of each, and the manufacturing system can cope. Military machinery – the 101in vee-8-powered machine, the lightweight '½-ton' Land-Rovers and more normal looking 109in vehicles – have their own special line; though again, in urgent cases or during out-of-balance demand, civil Land-Rovers could be built here, and military '½-tons' or 109in '¾-tons' have been built on the regular lines.

Lastly, and taking pride of place because of its high-performance product, is one hard-working Range Rover line. The higher price and more demanding specification of the Range Rover does not mean a slow rate of production. Production planners have found enough space for the job to be done, for quality to be maintained, and for up to 300 a week to be assembled. Up to sixty Range Rovers are built every day, one dropping off the end of the line every 8 minutes or so of the shift – not, by any means the sort of rate normal at Longbridge or Cowley, but quite enough when one considers the special equipment used and the mechanical complexity of the product.

Military production does not run at the same pace, and there are Ministry of Defence inspection processes to be accommodated. In any case, in terms of Land-Rover engineering and fittings, the 101in forward-control device is substantially more complex than its fellows, and still a relative newcomer to the Solihull factory's assembly workers.

In certain places at Solihull there is a great deal of coming and going. Overhead slings arrive on rails, dangling half of a Land-Rover body, a diesel engine, or an axle unit; wheels and tyres rattle down a chute; or a fork-lift truck fusses up to the edge of the tracks with pallets of headlamps, wiper motors, trim panels or handbrakes. In other places nothing seems to be happening; but look inside the engine bay, in the driving compartment or even under the tail, and you will find skilful hands going about their business.

Each part-built Land-Rover has a little card tucked into a facia slot bearing a list, typed by a computer, of instructions defining the vehicle to be built, who is buying it, where it is to be delivered, its engine type, chassis length, body variant, standard extras, special extras, etc. This card is the key to Solihull's jigsaw, and has to be obeyed. Species, destination and special features have already been fixed, and now it is up to the fitters and supervisors to act as midwives at the birth. Rarely do they have problems.

The heart and soul of any machine is in its engines, but the muscle is always in its frame. So there, with the Land-Rover, is where we must begin. At Solihull, everything centres around the massive black skeletons. Once they begin to creep down the assembly lines, they are the masters. Activity builds up all around, and every other part has to be brought to them, offered up to them, and bolted into place.

For most versions, British Leyland make their own frames, near the centre of Birmingham, in Garrison Street. Under the looming walls and alongside the uneven yards of a former Victorian rope-making factory, the frames begin to take shape. Welding machines, fixtures, anti-glare curtains and paint-dip baths have taken the place of old-style engineering. When Rover took the place over in 1965, massive reconstruction was needed. Today, on three floors, nearly 800 people are building chassis frames, welding doors, assembling dash panels, and stitching up seats. Garrison Street is one of the several vitally important 'satellites' on which Solihull depends for so much.

One end of the building is piled with flat steel sheet – some straight strip, some profiled – and completed side-members. The company welds up everything for its 88in frames, but 109in side-members are 'bought in' already complete.

As soon as the process begins, one steps back mentally to 1948, when Olaf Poppe and Gordon Bashford developed their cheap and effective assembly methods – which are still very quick and effective.

Four long steel strips, two straight and two profiled are slotted into the first fixture, tack welded, and swung round to the next machine, where the continuous welding process still operates. All of a sudden a recognisable bit of Land-Rover appears. At the next station the first cross-members are added, and the chassis frame begins to take shape. For the next hour or so there is a confusion of bright flashes, welding guns swinging to and fro on balance springs, men in leather coveralls wearing thick goggles, and an eerie progress in and out of closed-off welding booths. Some hours later the finished job sits on its trolley, ready for the paintshop staff to take over.

To the knowledgeable, most frames are recognisable, for there are vital differences. Men on the track talk of left-handers and right-handers, WDs and a mysterious hybrid called a 'composite'. There are differences between civilian and military frames, but only in detail. CKD frames have many cross-members and brackets only lightly tacked into place, 'to save space in packing, and to give local assemblers something else to do'. Once a CKD frame was sent to Solihull for assembly, and it was used for a vehicle, 'but within a hundred miles the driver was back to complain that all the body mountings had fallen off!'

Like Solihull, Garrison Street is crammed full of work. At one time, in the depths of the economic gloom after the Ryder Report was published, it was the only BLMC factory in Birmingham that was still recruiting new labour.

The 88in Land-Rover flown in to help the Range Rovers in the Darien Gap had a very hard time. Every panel is battered, and it has been capsized twice already

A one-off special, actually a Series IIA, towing a train load of other IIAs

This hovercraft conversion even kept the wheels clear of the ground, especially useful for crossing seeded land

A Dixon-Bate conversion to short chassis and articulated connection

Range Rover and the very special 101in military frames are bought from an outside supplier. Every working day a dozen lorry loads of chassis are sent to Solihull, where they join other important parts made outside the parent factory. Vee-8 engines are machined and assembled at Acocks Green, from components made all over the Midlands. Four-cylinder petrol and diesel engines, and the six-cylinder petrol engine, are all made at Tyseley No 1.

Machining, assembly and testing of Land-Rover gearboxes and transfer boxes takes place at Percy Road – its sole function. Front and rear axles are machined and assembled at Perry Barr, north of the city centre, along with similar components for the Range Rover. Perry Barr specialises in transmissions for the Rover company, but Range Rover transmissions take shape at Tyburn Road, Erdington, in the shadow of Spaghetti Junction, and are then transported to the Tyseley 2 factory for assembly.

At Clay Lane, Coventry, a small labour force concentrates on cabs, tropical roofs and the familiar hardtops for Land-Rovers. All body pressings, apart from the dash panel and a few minor details that are made of corrosion-proof light alloy, are stamped at Solihull, and bodies begin to take shape very close to the press shop and final assembly lines.

Special fittings come from carefully selected outside suppliers; there are hundreds of extras and dozens of suppliers. Incidentally, Rover rarely build an absolutely standard Land-Rover, without extras. A production man told me: 'I can't remember one at all in recent years. A Land-Rover is better for its extras anyway. By the way, if we set out to build a sequence of Land-Rovers all of which were marginally different from the others, it would take years before we had to repeat ourselves.'

The Land-Rover is one of the most important of all British Leyland's products, but in spite of its dollar-earning potential, and its continuing success, it is a long way from the immediate attention of British Leyland chiefs. This family tree illustrates the point: Plant Director for the 4 × 4s, alongside his Rover private-car colleague, is Alec Mackie; his boss is I. K. Showan, Managing Director of the Body and Assembly Division, who reports to Bill Davis, in charge of every British Leyland manufacturing operation; Davis is one of a handful of men with direct access to Derek Whittaker, Managing Director of Leyland Cars; and Whittaker is on the main Board of Directors. Whittaker only needs to concern himself with Land-Rover affairs when something goes drastically wrong. When production continues relatively smoothly, Alec Mackie can get on with his job, but at the first signs of a serious stoppage, the Managing Director will be on the phone to ask why.

Land-Rovers have been rolling out of the same Solihull buildings since 1948, and it would be easy to think of their production as a self-generating operation. Except for one or two very isolated instances, sales have always been buoyant, and ahead of the ability to deliver. It takes a long memory to pinpoint the last time production kept step with orders.

Just before he retired, after more than a quarter of a century's service in the company, Bill Mason, the Land-Rover plant's Materials Control manager, talked about his problems. It was his job to ensure that the right sort of components, in the right quantities, were always available, and at the side of the track.

> We call each of the 'satellite' factories 'suppliers', even though they are in the group. My people schedule on them, and on hundreds of independent suppliers, everything – every nut, bolt and washer – that goes into making Land-Rovers. Some time ago we switched over to mechanised scheduling, which means that a computer aids us in this. I am responsible for material arriving in good time alongside assembly lines, to a build plan which my people have already produced.

Even with a computer to help, 500 people are needed to run Materials Control. A minimum stock needed to keep the lines running smoothly probably means £8 million of parts in stock at Solihull alone. So how far ahead did Bill Mason have to start scheduling things from distant locations?

It can be as long as six months before I need them here. If I can explain that in more detail – we schedule roughly two months' firm needs, week by week, then a further two months tentatively (and we reserve the right to amend that part) and finally a two month forecast to allow suppliers to go ahead with material procurement. This applies to inside *and* outside suppliers.

There are elaborate safeguards against shortages, stoppages and transport problems. Normal components, used in large numbers, might be held with a minimum of seven working days of stock, but for bulky and expensive items like shock absorbers, road wheels and steering gear, there is no space for more than five days' needs, and tyres can be stocked for only two days' requirements.

If there was space at Solihull for increasing production, Mason believes the supply system could cope with an immediate 10 per cent increase at once: 'We could build 1,350 Land-Rovers one week, and nearly 1,500 the next, without running out of parts. The system's that flexible – always assuming, of course, that the "mix" remained about the same.' If the objective was not an increase in production, but a change in 'mix', perhaps to give more emphasis to the longer-wheelbase models, he thought that something like a 10 per cent swing could be obtained within a week. More radical changes would take time. A change from one basic model to another would be difficult: it would not be feasible, for example, to build Range Rovers on the military 101in lines.

> The problem is not so much one of space, as of equipment, overhead conveyors, body access to a new mating station – things like that.
> We have 88in lines and 109in lines, and we try not to mix the two – not if we can help it. We can put 88in vehicles down the station-wagon line just like that, but we can't put them down the 109in line just like that.

The problem in ensuring supply is not always in having enough of anything, but of not having too much. For the Land-Rover, there are thousands of components, coming in by transporter from at least 700 suppliers.

The production sequence starts with a 'sanction' from the directors to build a given quantity, in thousands or tens of thousands, of a given model. In that sanction the sales staff will know what proportion of each important variant should be produced; from their own experience and from computer forecasts they can pick up and allow for changing fashions.

Sometimes the Production Control staff may object that a proposed sales programme would result in inefficient production and hold-ups. If, for instance, too many Range Rovers were requested with the optional power steering (a fitment needing extra time to link up to the vehicle itself), it might affect line speeds and therefore output. But the problems diminish with time. Experience, and a long-running model, have sorted out most annoying wrinkles by now.

Rover do not build vehicles for stock. Mason says: 'Every Land-Rover is ordered by a distributor before we start on it. Every one has a "sales label", and at any one time we hold 4,000 of them. From that bank we compose build schedules for seven working days. My people are experts. They distribute the orders including time-consuming extras among more basic build requests – that way we always keep the tracks moving.'

Supplying the moving tracks is a mammoth job, though, since the rest of the motor industry is currently in the doldrums, the problems have eased during 1975. Buying expertise now includes knowing when a supplier's next important wage negotiations are due. It is no credit to a buyer to ignore possible stoppages, and the more experienced can tailor their orders to that very possibility.

Of the 700-odd suppliers of bits and pieces, some stand out because of the money spent with them and the size of their deliveries. It illustrates the complex nature of the British motor industry to list them:

GKN, which includes Qualcast Foundries castings, sheet alloy from Birmingham Aluminium Co, and cylinder blocks from Midland Motor Cylinder

Triplex, for glass used in windscreens and other windows

Dunlop, for tyres and other rubber goods

Rubery Owen, for heavy pressings and chassis items like wheels

Lucas, which includes CAV and Girling, for electrics, brakes and hydraulic details

Burman, for steering gear

Smiths Industries, for instruments

Jonas Woodhead, for road springs and shock absorbers

British Steel Corporation, for sheet steel used in chassis frames and body construction.

Land-Rover production rarely falters through lack of supply, and when you consider that a Land-Rover 'sales label' is usually liberally covered with extras, that is a remarkable achievement. There are rarely fewer than ten on the list, and often over twenty. The 'record' stands at no less than eighty!

Extras call the tune on production. If, for instance, there was a massive hold-up on one particular type of engine, the lines would have to concentrate on Land-Rovers with different types of engine. Almost certainly this would alter the balance of 'call up' for extras, and might lead either to shortages or to a spot of hard bargaining with the sales division.

If you are ordering a Land-Rover personally, you would be well advised not to change your mind on equipment at the last minute. A winch instead of a power take-off might be easy to accommodate, but probably not a diesel instead of a petrol engine. If the worst came to the worst, you might 'lose your place' in the queue for your Land-Rover to be built, and it is a very long queue. It takes about a day from the moment the first item is bolted on to a chassis frame at the west end of the Solihull works to the moment when the completed Land-Rover is driven off the main assembly line. It might take up to a week for the vehicle then to go through the finishing and final lines, and undergo any rectification that might be needed.

A Range Rover, built in only one basic form, takes shape rather more quickly. It still takes a day for the skeleton to become a running machine, but the process might be complete inside three days.

After that the distribution network takes over, and soon the vehicle will be all yours. Rover started scheduling its build 6 months earlier, and took a week actually to do the job. If they have kept up to their usual high standards, it ought to last for 10 to 20 years.

7

ROAD ROVER TO RANGE ROVER

No other 4 × 4 vehicle anywhere in the world combines the same virtues as the Range Rover. Rover hit the jackpot in 1948, and they did it again in 1970, when the Range Rover, like the earlier Land-Rover, started a new trend and left the rest of the world to struggle along behind.

A Range Rover is no Land-Rover. For one thing its price puts it in a completely different league, as does its performance. Even so, the Land-Rover design team can claim it as one of their own, as the badge on the car's tailgate testifies, and much of the design philosophy is the same. If you consider the latest military Land-Rover – the 101in wheelbase vehicle – and notice how Range Rover engine and transmissions have been fitted without change, you will see how one layout feeds from the other.

The most remarkable thing about the Range Rover is that it took so long to evolve. Almost as soon as the big vee-8-engined machine had appeared there was a Press chorus of 'Why haven't Rover done this before?' In fact Rover wanted to, many times, but somehow there were always more pressing matters to be settled first.

The Range Rover, with its beautifully integrated design, cannot be considered in isolation. It was not finalised until the late 1960s, but that does not mean that its concept was new. Like many other things at Rover, new models and new ideas take ages to crystallise, and the Range Rover that we can buy is not the machine it might have been. This is hardly surprising, for it was first conceived in 1950 or 1951 by Maurice Wilks, the 'father' of the Land-Rover, and the creator of the long-running P4 passenger cars. However, as with any new Rover design started since World War II, he turned to Gordon Bashford for help. There is an unwritten law at Solihull that 'Bashford always does the first schemes', and this was no exception. He recalls: 'Maurice Wilks had decided that we now needed a new vehicle, still with cross-country capability, that was more suitable for road use, as opposed to the workhorse concept of our Land-Rovers. On that basis I was asked to work up the first layout, or package. At that time it was to be based on P4 components.'

P4 was the general name for Rover's latest range of saloon cars, conceived in 1946 and 1947 by Bashford and Maurice Wilks, and trading on all the experimental work carried out at Chesford Grange immediately after the end of the war. The P4 – or the Rover 75s, 90s, 105s, 80s, and 110s, as they were called in the showrooms – was a very staid and gentle car, nicknamed 'Auntie', a name that stuck, even inside Solihull; and it seemed at first that nothing could have been less suitable for the basis of a new composite vehicle. However, the P4s were all based on very rigid chassis frames, the engines and transmissions were new, and they were already building up an enviable record for reliability. That was the engineering justification. The practical justification was that even the prosperous Rover company could not afford yet more new tooling, and would have to settle for existing parts. Gordon Bashford says:

> We messed about, scheming layouts for two- *and* four-wheel-drive at first. Even though the P4 chassis was a strictly conventional car component, we could still use standard parts and ingenuity. We went through a whole series of schemes, and I promise you that I also considered front-wheel drive as well!

There is nothing unique in this. Gordon Bashford dabbled with front-wheel drive schemes many times. He considered it for the Rover 2000, and even for the stillborn 4-litre vee-8 P8 project;

This Series III Land-Rover seats up to twelve people

Special conversion to a forward-control Land-Rover for an export customer

Long-wheelbase Series IIA military Land-Rover towing a 105mm howitzer

The first of the Road Rovers, nicknamed 'The Greenhouse'. Underneath this austere body was a P4 Rover 75 chassis and the 2-litre Land-Rover engine. This was built in the early 1950s

Front-wheel drive transporter for the Rover-BRM Le Mans car

The first King-Bashford-styled prototype '100in station wagon' of 1967. It is really remarkably like the final production shape

Solihull's production version of the Range Rover. Compare this with the prototype on previous plate

Range Rovers have a lot of wheel movement and the ground clearance is exceptional

and had it built into the T4 gas-turbine car, which used a Rover 2000's structure, where it worked very well.

> The original prototype was, in fact, four-wheel-drive, with a lot of Land-Rover parts, though we had to chop about the chassis a lot to achieve this. Maurice Wilks had come up with another new name – the Road Rover – which was perfect, and told you everything. But that first car looked very odd and we all called it The Greenhouse. It really did look like one too! It was more or less styled by Maurice Wilks.

Maurice Wilks's styling ideas were much influenced by what the Americans were doing at the time. He had had a very sure eye for a line when the classic 1930s Rovers were shaped, but he was not at home with all-enveloping shapes; the fact that he set up a specialist styling department in the 1950s meant that he had recognised his own limitations. Even so, for several years the youthful David Bache had a very difficult job to do, where diplomacy counted for a lot.

Maurice Wilks announced his intentions to his co-directors in 1952, and they came to a tentative agreement that the car should be put into production in 1953. Work was to carry on for several more years, however, and the specification of the car to be changed quite radically and often. Yet it was never seen in public. Gordon Bashford explains why:

> Everybody seemed to like the first car, and after the directors had approved it the vehicle was moved into Dick Oxley's area. This meant that it had become a current project – my department dealt with basic research. Despite Dick's efforts to maintain the original, simple, slab-sided concept on Land-Rover lines, his brief from management escalated. They thought the original was too austere, and asked for changes to the light-alloy body shell, involving the shape and needing complicated pressings. In this way it got bigger, grander, heavier and more costly.
>
> It never really had much priority, and by the time all the modifications had been done, Land-Rover station-wagon production had built up so far that it didn't look viable.
>
> Of course, also, the Road Rover was in no way a cross-country

vehicle. With conventional rear-wheel-drive it couldn't have been. It was initially meant to be only an austere station wagon – in the end it wasn't all that austere any more.

By 1957, with production tooling still not wholeheartedly committed, the Road Rover had the following specification. Although the basic P4 passenger car chassis was used, its wheel-base was chopped from 111in to 97in, while the main side-members were narrowed by ¼in. Front and rear suspensions were like the P4's, but modified in detail, for the designers had decided to use independent front suspension with laminated torsion bars. The steering gear was mounted well forward in the chassis, so that there was space for future optional engines to be installed.

The engine itself was to be the 1,997cc four-cylinder Rover 60 unit, which was also fitted to Land-Rovers at the time. If the Road Rover had reached its customers, the standard engine would almost certainly have been the ohv 2,286cc unit, which Land-Rovers inherited in 1958. There was to be no optional engine at first, though as prototype work had been done with the six-cylinder 2,638cc Rover 90 engine, and space existed in the engine bay, this was a distinct possibility.

The usual private-car gearbox was fitted, a Laycock overdrive would have been optional, and the gearing was like that of the saloons. Even the tyres might have been the same, though over-size 'knobblies' were to be optional.

The body itself, recently restyled and looking very much like a current Chevrolet wagon, would have had a two-door shell, with upper and lower opening tailgates. The dash panel, door inner frames and a few other details would have been pressed from steel sheet, but the rest, including all the skins, would have been pressed from Birmabright.

In spite of the light-alloy body panels, the Road Rover ready for the road with its 2-litre engine weighed about 2,850lb, which seems heavy until one remembers that the P4 passenger car from which it was derived weighed 3,300lb and provided much less space. It was not too large, either: the overall length was 13ft 8in, the width 5ft 3in and the height 5ft 3in. The ground clearance was a mere 5·4in, which would have precluded any cross-country travel the management might have had in mind.

Spencer and Maurice Wilks were still interested in the type of car the Road Rover represented, but they could not bring themselves to put it into production. With more time to spare, and if dramatic developments like the all-new Rover 2000 project had not been brewing, they might have worked another miracle. A Board meeting in March 1958 administered the *coup de grâce*: 'In view of other commitments it had been found necessary to postpone the introduction of the Road Rover.'

The Road Rover project had taken a long time to die, but the idea of the car was not forgotten. Even so, it would be some years before Peter Wilks or Spen King could get round to that sort of concept again. But always, nagging away at the back of their minds, was a conviction that perhaps the world *would* buy a lot of 'luxury Land-Rovers'. By the middle of the 1960s Spen King found some spare time. He had shrugged off his day-to-day worries about the Rover 2000, now safely in production. His beloved gas-turbine cars had also been side-lined, and his ideas about a new large car to replace the old P5 3-litre saloon were still at an early stage. It was the perfect moment to look at four-wheel drive vehicles again.

There were all manner of portents. In North America the big Jeep Waggoneers had shown that a successful cross-country machine did not have to be stark and uncomfortable; from Japan (and this was hurting most of all) came Toyotas, which combined Land-Rover agility with much higher trim and furnishing standards; and a BBC *Wheelbase* programme had heavily criticised the Land-Rover, even though it was still demonstrably a worldwide success.

Once planning had started on the new machine, it was agreed that it could have little in common with the Land-Rover. Spen King, as usual, enlisted Gordon Bashford to be his co-thinker

113

and right-hand man on the project. Both were convinced that they could not provide any sort of luxury if they retained the Land-Rover's hard and rugged suspensions. What they wanted (and this will certainly be on Spen King's epitaph one day!) was much more wheel movement, space, and refinement. These could not be got from the 88in or the 109in Land-Rover, so a radically new vehicle was needed. In any case, even though King and Bashford were agreed about four-wheel drive, the new concept was quite new; the Land-Rover was a cross-country vehicle that could be used on the roads, but the new design would be a road car that could cross unmade terrain if necessary. The new machine would be much more like a private car than its ancestors.

This concept made the next phase easy. It would be possible to start from the almost mythical 'clean sheet of paper', a rare pleasure for any engineer. Even though the designers would have to accept one of Rover's existing engines, the rest – transmissions, engineering, packaging and proving – was all up to them. Gordon Bashford still recalls the way he started the design:

> We started off in the early 1960s by considering the car with the old six-cylinder P5 3-litre engine. We called it the '100 in Station Wagon' because when I had finished sketching up the first package, the wheelbase turned out to be 99·9 in, so I said we should round it up, and call it after that dimension. That was one of the few things we didn't change from first to last. We had no serious competitor in view at that time, because I don't think there was an existing machine which did what we were trying to do.
>
> In order to get a good ride and acceptable performance across country, an awful lot of thought had to go into the suspension design. It was then that we decided that we must have low spring rates, large wheel movement, and good damping. Self-levelling (which we have at the rear) came in at the same time. If you have low rates and long wheel movements with a high payload potential, then levelling is essential. That wasn't going to be easy, but Spen King and I visited the Frankfurt Motor Show that year where, lo and behold, we saw the Boge Hydromat levelling strut. It was ideal for what we needed. It wasn't a damper and it didn't need extra power – it powered itself, pumping itself up to the

pre-set level as the car started off up the road. Reliable too –
Mercedes were using it on their passenger cars.

Even as first conceived, the 'rooin' would have been a good
car, but possibly not dramatically so. Something was missing.
But that arrived a little later – Rover's new vee-8 engine! There
is no doubt that the Range Rover with its glorious, powerful
light-alloy power unit is twice the machine it might have been;
and although petrol consumption can sometimes be heavy, as
independent magazine tests have shown, the engine is one
feature that really 'sells' the car.

A vee-8 engine, however, after years of slow progress at Rover,
may seem surprising. The credit must go to ex-Managing
Director William Martin-Hurst, and it is his story:

> I was on a visit to Mercury Marine in North America to talk
> Carl Keikhaefer into buying Rover gas-turbine engines for his
> pleasure boats. However, by the time he and his chief engineer
> Charlie Strang had studied all the details, documents and balance
> sheets relating to Rover, Strang suddenly said, 'I see you have a
> 2¼-litre diesel', which was, of course, the Land-Rover diesel. He
> then told me how they were developing an inboard-outboard
> scheme for Chinese fishermen and that they were using Mercedes
> diesels. So I supplied him with a couple of Land-Rover diesel
> engines, which he fitted into boats down at his private lake in
> Florida.

That might have resulted in very substantial business for
Rover, but in the event it led to more sensational developments.

> One day I was in his experimental workshops in Fond du Lac, in
> Wisconsin, talking about this and that, when I saw that lovely
> little light-alloy vee-8 engine sitting on the floor. I said, 'Carl,
> what on earth is that?', and he told me it was for a racing boat, and
> that he'd winched it out of a Buick Skylark car. I asked him
> whether it would be available, and I was astounded when he told
> me that General Motors had just taken it out of production!

Martin-Hurst then ran a rule over it, and found it to be very
little longer than the Rover 2000's four-cylinder unit, and only a

few pounds heavier. As Rover were busily trying to slot six- and five-cylinder engines into the new Rover 2000 at the time, with expensive results, Martin-Hurst was intrigued at the possibilities for Rover, and decided to approach General Motors: 'I then had incredible difficulties finding the right man at GM to talk to, and I ended up having breakfast with Ed Rollert (who was at the New York Motor Show) to talk about it. For a long time nobody at GM would take me seriously – they didn't see why Rover, in England, wanted to use one of their cast-off designs!'

Negotiations took months, but at last Rover acquired their vee-8 production rights. The engine was redesigned to suit British methods, with the help of a GM engine designer near retirement, who was flown to Solihull, installed in a flat, and used as a design consultant. As far as the Range Rover was concerned, it had found its engine, and its prospects were transformed.

That vee-8 engine found its way into the stately old P5 3-litre car, into the P6 (as the Rover 3500), and even into the Morgan sports car. It would also have been used in Spen King's masterpiece, the P8 saloon, if British Leyland had not killed it off as too good to compete with Jaguars, and it is one of several engines to be offered in Rover's brand new 1976 saloon car. In every way the ex-Buick vee-8 3,528cc engine has become a cornerstone of the Rover structure.

Such was the King-Bashford reputation in new vehicle concepts that they were allowed, even encouraged, to style their new creation. As they meant the car to be a practical estate car, with lots of ground clearance and a permanent four-wheel drive for cross-country going, many dimensions were already fixed for them, and their efforts were remarkably successful. Rover's Styling department, under David Bache, were most complimentary about the prototype. In any case, in the mid-1960s, the department was far too busy to take on another new project, for they were engaged in retouching the old P5 3-litre, preparing the Rover-BRM Le Mans car, and starting on the very important P8 saloon.

Externally the Range Rover, as we know it, is recognisably the same as that prototype built in 1967. When it was handed over to David Bache's office, much attention was given to detail construction, to fittings and to 'produceability'. The facia and controls, too, were a Styling speciality. Even so, the general shape, the layout and the styling masses were left strictly alone.

All the mechanicals, of course, were new. Engine, main gearbox and the four-wheel drive layout would be peculiar to this car. Tom Barton's engineers had only their lengthy experience to fall back on when tackling the job.

The one big change in philosophy was that the new machine would have permanent four-wheel drive. Then, as now, people could not understand why a more luxurious machine, likely to spend less of its time 'off the road', should have this feature, when on the more rugged Land-Rover it had to be 'plugged in' manually. It was rather a marketing somersault. The Land-Rover, of course, had an 'optional' four-wheel drive, and matchless cross-country performance; the Road Rover would have had rear-wheel drive without the option; and now the new machine was being given four-wheel drive, also without the option.

Rover planners refused to accept that this was confusing. They never had qualms, and alternatives were never seriously considered. The reasons for their choice lay in the pricing, and in the type of machine they thought the Range Rover should be. However, the simple type of four-wheel drive layout found in every Land-Rover was not refined enough for the new car. To take care of transmission wind-up, and to improve traction even further, the new car was to have a third, central differential, with a limited-slip mechanism inside it (now deleted). That allowed the removal of one of the crop of gear levers in the driving compartment, though one was still required to select the low-range gears, essential for really exceptional climbing.

Apart from the new power train, there was plenty of technical innovation to keep the press happy. Disc brakes on all four wheels, self-levelling rear suspension, a live front axle where independent suspension might have been expected, safety belts built in to the front seats, and a host of other details all added to the interest.

Prospects for the new car were so encouraging that its release was rushed forward by the new British Leyland management. Rover managers, unaccustomed to haste when it came to capital commitments, might have taken more time if the firm had still been independent. The first prototype was not finished until August 1967, several months after Rover had merged with Leyland-Triumph, but the first production car was shown to its public in June 1970. This was not dramatic progress by comparison with the first Land-Rover, which made the same jump in about one year, but by current Rover car standards it was breakneck speed. To experienced observers of the motoring scene, the rush was obvious. Deliveries started very slowly and waiting lists built up rapidly. Once deliveries began, a new legend began to form. Surely Rover had done it again? This was another versatile machine that could go on for a generation.

The customers had to decide for themselves what sort of machine it was, as they had done with the Land-Rover. Rover's management were sure of its engineering, sure of its versatility, but unsure of its new clientele.

At first they had to sit back and wait for trends to develop. They had to find out whether the Range Rover would sell to cross-country users, to the well-to-do, to the 'leisure and pleasure' market; and whether it would be used as a workhorse or as a large and commodious estate car. With the original Land-Rover, of course, they had guessed wrongly. They had intended it to be a tractor that could be used on the road, but its customers had found other uses for it. What about the Range Rover?

In a survey carried out in 1972 Rover distributors in Britain persuaded several hundred Range Rover customers to list their professions, and their plans for their new buy. Nearly two-thirds of the vehicles were being bought for dual-purpose use, which scotched the theory that the Range Rover was too elegant to be

With tracks to each wheel, this Land-Rover was just about unstoppable.
The army used it to help in bomb disposal

Land-Rover station wagon wallowing in deep water on the Eastnor estate, where much Land-Rover development is carried out.

taken off the roads, and the vast majority of the customers were engaged in business, farming, estate management, building or construction. Only three purchasers stated that their Range Rovers would be used purely for business, and of the rest, the purely 'private' customers, most were in the director/senior executive/professional classes, and would use them for leisure and pleasure. More than 70 per cent of the respondents said they would be using Range Rovers for towing such loads as horse-boxes, caravans, boats and trailers; and nearly half said that their new machines would not normally be used 'off the road', except for manoeuvring and in emergencies.

The bad news was that about one in three of these customers were trading in Land-Rovers or Rover cars, and said they would normally have made a direct exchange. In motor industry terms this was 'substitutional selling', something any firm would rather do without, for it is not good for any new model if its sales are achieved at the expense of existing cars in the same maker's range. This figure, however, was falling all the time, and was expected to continue falling.

Peter Garnier, then Editor of *Autocar*, soon found out what the Range Rover was not:

> I found myself wondering exactly what market my Range Rover appeals to. When I first took it over, I made a point of showing it to several farmer friends at home.
>
> 'I've got *just* the car for you', I said proudly, 'It'll do everything you want – run around the fields all day, loaded up with hay, cattle-feed, pigs, anything you like. And then you can put on your glad rags and go out to dinner in it in the evening.'
>
> Invariably the answer was: 'You're wrong. We use our Land-Rovers all day for the job, and the only cleaning they get is when the muck starts falling off them. We want something entirely separate in the evenings – to forget work and go out in our 2·5 PIs and Volvos. Who wants to go to a party in something that smells of pigs?' I could understand the argument.

What the Range Rover has, however, is a presence, an ambience, and a touch of authority. On the very day the car was announced, a friend arrived at my house in a press car version, and we went out for a spin. First impression was of the magnificent view from 'the bridge'. A Range Rover is 5ft 10in high, much more lofty than most modern cars, and I had forgotten how much of the countryside is no longer visible from them. The second impression was of the way other traffic reacted. We would drift up behind another car, then toot and wait for the reaction. The driver, glancing in the mirror of his car and seeing the massive Range Rover nose behind him, would soon make way.

Range Rovers have only been on sale since 1970, and have already doubled in price, but sales continue to rise, that single Solihull assembly line is full to bursting, and the waiting lists persist. The army has not found a use for them yet, but the police and ambulance services have taken them to their hearts. There is little more impressive than a motorway Range Rover patrol car, with blue lights flashing, steaming up the outside lane at anything up to 100mph.

The Range Rover won Coachwork Gold Medals at Earls Court in 1970 and 1971, in addition to the Don Safety Trophy for its engineering features, and, in 1971, the RAC's Dewar Trophy, Britain's highest automotive award, for 'advanced development in automobile design'.

There were a few complaints, which were speedily dealt with. The steering was a touch heavy for a woman driver to haul around, and power steering therefore became optional. The rear window could become coated with dirt, so a wipe/wash feature is now standard. Trim and furnishings were thought to be somewhat spartan, and the latest specification, with its carpeted floors, is rather more comfortable. Instrumentation is now comprehensive, to take care of the original criticism that it was too meagre; and dogs that could not keep their footing on the untrimmed rear deck are now able to stand comfortably on a moulded floor mat.

When the Range Rover's British selling price broke through the £4,000 barrier in 1975, some Rover managers were a little apprehensive. No Rover car had been so expensive before. Would it be too much for the customer? It was not. Sales have continued to mount, and more than 10,000 were built in a year for the first time.

When the Land-Rover started, it was just one machine, but now there are dozens of variants. Will the same thing happen to the Range Rover? Will there, for instance, be the rear-wheel drive, road-tyred version, which is all some customers need, a long-wheelbase variant with four passenger doors perhaps, or an open-topped vehicle? These may come in time, but at the moment British Leyland have enough difficulty satisfying the existing demand.

APPENDICES

APPENDIX 1

THE ROVER COMPANY

When the Wilks brothers decided to build Land-Rovers in 1947 they could have had no real idea what they were starting. The 'useful stop-gap' soon became dominant at Solihull, and – before long – the company came to rely on it. Financially speaking, within a couple of years – no more – the Land-Rover had completely changed the company's prospects.

During the 1930s Spencer Wilks set out to mould Rover into a new shape. He envisaged a compact, profitable, medium-sized and respected concern, which would make as many of the type of cars he preferred as he decided to make. They would be advertised and sold in a discreet and orderly manner, and built at a civilised pace. The cars would be thoroughly sound, well-engineered, high-quality products.

This policy, of course, was a huge success, and by 1939 Rover had a splendid reputation. The firm was the envy of its rivals. Financial worries there might have been during 1931 and 1932, and in June 1932, in fact, Rover and Triumph had discussed merging, but by 1939 Rover was prosperous and superbly managed.

Even World War II, the running of two large 'shadow' factories, and a change in future prospects, could not alter everything at once. Rover was still a particularly safe and soberly run concern. Indeed, by the end of the 1940s, Rover were already quite large enough to take over other companies, if Spencer Wilks had been interested in doing so.

Then came the Land-Rover, to transform the company and its forecasts. Just as soon as it became clear that Rover had a worldwide success on its hands – that here, properly cosseted, was a machine that could be the firm's profit-earner for the next generation – several business tycoons sat up and began to take notice.

They might not originally have been aroused at the thought of owning a 200-a-week car business, but the attraction of a best-selling and near-unique cross-country vehicle was something else. There were a few drawbacks, of course. Rover, for instance, were pressing ahead with research into gas-turbine engines for boats and cars, which most rivals were convinced was misguided; but gas turbines were for the future, and research spending could always be cut out. The Land-Rover was here and now, hard evidence of success.

It was natural, then, that Rover talked mergers, associations, links, etc, with several other concerns. There was rarely a year during the 1950s or 1960s when the directors were not in discussion with one prospective partner or other. Not that the company was ever financially in trouble – far from it. Any talks were initiated from the base of strength, for Spencer Wilks was never interested in joining forces with anyone as a junior partner.

The most public discussions, and the most persistent, were with Standard-Triumph. These had nothing to do with the relationship between Spencer Wilks and Standard's Sir John Black, who were brothers-in-law, each having married a daughter of the motor industry pioneer William Hillman, for there was no formal inter-company contact until after Sir John had resigned from Standard-Triumph in January 1954. Later that year, in the spring, Wilks made the first approaches to Standard's new managing director, Alick Dick, who immediately saw the attractions: Rover had the Land-Rover, while Standard were building tractors on behalf of Harry Ferguson; Dick knew that Rover wanted a diesel engine option for their Land-Rover, and that he could perhaps supply it. Rover were also attracted by the prospect of spare manufacturing capacity in Coventry, which they already urgently needed, and by the likelihood that Rover and Standard-Triumph car ranges might dovetail quite logically. Equally as important was that Wilks knew Willys-Overland of the United States had been looking for a European manufacturing base for some time, and that they had already been talking to Standard. What also came out of the original round of talks was that Standard were considering making a four-wheel drive vehicle of their own.

In 1954 everything speedily fell through, not because a merger would have failed in practical or financial terms, but because there were personal disagreements. Rover were concerned about Standard's less impressive money-making record, and wanted to be dominant in future planning; whereas Alick Dick and Standard-Triumph wanted to be equal partners at least. Technically, Maurice Wilks' team was most disappointed when they found that neither the Standard diesel nor the Vanguard petrol engine would be suitable for use in the Land-Rover. One of the dying attempts to make a merger feasible was to consider the joint design of new engines both for Ferguson tractors and for Land-Rovers.

Five years later, in 1959, when Standard had sold off their tractor interests to Massey-Harris, and were momentarily stuffed with cash, Rover decided to talk mergers once again. This time there was more emphasis on an interlocking of the new car programme, though the established success of the Land-Rover, now with a diesel engine of its own, was a big factor. Rover learned that although Standard had not carried on with their own four-wheel drive projects in the mid-1950s, they were now considering the manufacture of a tractor of their own design. Once again the equation 'Tractor + Land-Rover = profit' was considered. At the same time each company discovered that the other was developing a 2-litre 'executive' saloon car, and the competition, if they stayed separate, would be intense. But once again the talks collapsed, probably for the same reasons as before, and Rover would not finally be linked with Standard-Triumph again until 1967.

Rover were always worried about Willys, their Jeeps, and their rumoured interest in setting up a European manufacturing plant. The Land-Rover might have been more versatile than the Jeep but the latter had the glamour of its universal use during World War II behind it, and the possible marketing might of a large concern to boost its sales wherever necessary. Surprisingly, Willys do not seem to have made any formal approach to Rover at any time, though there was one discussion in 1958 that was more concerned with technical collaboration and distribution in overseas markets than with any thought of mergers.

The companies were each rather jealous of the other's products, and predictably there were no fruitful results. The question of export networks worried Rover in the 1940s and 1950s, for up to then they had only exported Rover cars to 'the British Empire' – and only a handful at that. With the Land-Rover, however, they were having to learn fast.

Rover engineers assure me that they never seriously considered making a tractor to back up the Land-Rover – they had made a tractor once, in 1931, but only one prototype was built – but from time to time they considered joining forces with other tractor manufacturers. Harry Ferguson was the obvious candidate, even if his tractors were being made by Standard-Triumph, and Spencer Wilks had informal talks with him during 1952 to discuss 'technical collaboration'. The 'grey Fergie' and the original Land-Rovers were already familiar partners on any progressive British farm, and it would have been very cosy to see them closer together, if only by sharing components. But with Standard-Triumph firmly entrenched as licensed manufacturers of Ferguson tractors, and with Wilks discovering the difficulties of dealing with Ferguson, there could be little progress.

There was obviously something magnetic about links with tractors, and in 1957 Rover were tempted again. This time it was an offer from the David Brown Corporation, which were interested in selling off their complete tractor and agricultural division. Rover found out as talks progressed that David Brown had designed, then abandoned, a 'Land-Rover' of their own; and, as Tom Barton has said, they had also tempted Rover engineers away to develop it for them. For a time this offer had its attractions, but Rover could not escape from the suspicion that David Brown were proposing to sell them a division that was not making profits. Talks continued in a desultory manner for some time before they were abandoned, and then they were reopened at the end of the 1950s. They were finally abandoned early in 1961, when the directors and the company were much too busy with the new Rover 2000 project to bother with anything else.

Discussion with the Austrian Steyr-Daimler-Puch company, which make the nimble little Haflingers, took place in 1962, but nothing came of it. However, it showed that the Rover board had come to terms with their cross-country marvel, and were looking at every possible way of expanding their interests.

With demand for the Land-Rover always exceeding supply, the shortage of production space had become critical by the early 1960s. Therefore, when Alvis approached the Rover company in 1965 and offered themselves for take-over, they were received with enthusiasm. This would not be a marriage of models, or even of philosophies – it would be a merging of skills. If Rover had to absorb any independent manufacturer, Alvis would be ideal. They had ample modern factory space in Coventry, a prestige car in the Park Ward-bodied 3-litre, and a range of successful military vehicles – the Saracen, Saladin and Salamander six-wheel drive machines. Unlike all previous merger talks, these occupied only a few days, and by July Alvis were under Rover's control.

There were two more corporate moves in store for the company. By the mid-1960s Rover directors were looking hard at their forward programme, and wondering how the new capital for their P8 car, their 100in station wagon, and the expansion of Land-Rover production facilities should be raised. An approach from Leyland Motors, therefore, at the end of 1966 was quite opportune, and shortly the Rover company, independent since its formation in Victorian days, agreed to be absorbed into the fast-expanding Lancashire-based concern. It was thus that Rover and Triumph finally got together, nearly 35 years after their original discussions. There were many logical, and even a few emotional, reasons why Leyland should acquire Rover, and there were good arguments for Rover staying aloof, for they were not in any financial trouble at the time; but with rocketing investment programmes ahead, it was not likely that they could have survived alone for long.

For the Land-Rover, this merger meant a new lease of life, clearing the way for the six-cylinder engine option to be announced, for its facilities to be somewhat modernised, and for a much-needed facelift to be carried out. For the Range Rover project the merger was life itself. Donald Stokes was reputedly so excited by the first Range Rover schemes he saw that he asked for its announcement to be brought forward. William Martin-Hurst confirms that the announcement was advanced by at least 12 months, even though his engineers pleaded for more time to finish the job.

Leyland were attracted by the Land-Rover, and made no secret of it. It was of most use to them as a marketing proposition in overseas territories, where their agents would now have an enormous range of commercial vehicles to sell, starting with the 88in Land-Rover and finishing off with whatever gigantic machine their Scammell division could deliver. Leyland did not propose to do anything drastic about Land-Rover production facilities, and since 1967, in spite of the great changes which have occurred in Leyland (later British Leyland), the vehicles have been built in the same place, by the same teams, under the same roofs. Functionally a Land-Rover might have more in common with a Leyland truck than with a Rover car, but that was no good reason to disturb Solihull.

British Leyland's financial problems, and the implementation of the Ryder Report, have caused enormous upheavals in the corporation but left Land-Rover and Range Rover activities untouched. For the engineers who designed them, the Report recommended that a corporate design facility should be set up – and Solihull should be chosen. This was a compliment to the place and the men in it.

Splitting up the activities into various categories has caused an amusing anomaly. Models in the Cars Division are now split into 'Small', 'Medium' and 'Large' sectors, and there was some difficulty in deciding whether the 4 × 4s were cars, trucks, or even Special Products. The result of these deliberations was a typical British compromise. Rover's Solihull plant, because of its cars, became a member of the 'Large Cars' division, along with Jaguar at Coventry, so that Land-Rovers, and Range Rovers are now designated Large Cars!

Whatever the Ryder Report might say about streamlining activities, it will make no sense to turn Solihull into an all-car plant. No matter what they might be called in the future, the famous 4 × 4s will be Rovers, and my guess is that they will only be built at Solihull.

APPENDIX 2

MAJOR MODELS, DATES AND SPECIFICATIONS

The millionth Land-Rover has little in common with the first one but its purpose. The machine has always been meant for the same rugged life, but has been progressively changed, improved, and modernised to keep up with the times. Between the first and the millionth there have been many important changes.

Tables 1 and 2 should clarify the evolution of the machine, and the text pinpoint the important milestones in its development.

TABLE 1

Evolution of the Land-Rover

Variants by wheelbase and engine			Years in production
S I	80in	1,595cc petrol	1948–51
S I	80in	1,997cc petrol	1952–4
S I	86in	1,997cc petrol	1954–6
S I	107in	1,997cc petrol	1954–8
S I	88in	1,997cc petrol	1956–8
S I	109in	1,997cc petrol	1956–8
S I	88in	2,052cc diesel	1957–8
S I	109in	2,052cc diesel	1957–8
S II	88in	2,286cc petrol	1958–61
S II	88in	2,052cc diesel	1958–61
S II	109in	2,286cc petrol	1958–61
S II	109in	2,052cc diesel	1958–61
S II A	88in	2,286cc petrol	1961–71
S II A	88in	2,286cc diesel	1961–71
S II A	109in	2,286cc petrol	1961–71
S II A	109in	2,286cc diesel	1961–71

133

	Variants by wheelbase and engine		*Years in production*
S II A	109in	2,625cc '6' petrol	1967–71
Army	88in ½-ton	2,286cc petrol	1968 on
S III	88in	2,286cc petrol	1971 on
S III	88in	2,286cc diesel	1971 on
S III	109in	2,286cc petrol	1971 on
S III	109in	2,286cc diesel	1971 on
S III	109in	2,625cc '6' petrol	1971 on

TABLE 2

Other variants

Wheelbase and engine		*Years in production*
Forward-control Land-Rover		
109in	2,286cc petrol	1962–6
110in	2,286cc petrol	1966–72
110in	2,286cc diesel	1966–72
110in	2,625cc '6' petrol	1966–72
101in	3,528cc 'V8' petrol	1975 on
Range Rover		
100in	3,528cc 'V8' petrol	1970 on

April 1948
The vehicle was first shown publicly. It made its debut on 30 April at the Amsterdam motor show. Only one version was available at first – the 80in wheelbase machine with pick-up body and 1,595cc petrol engine.

October 1948
A station-wagon version was offered for the first time, and made until early 1951.

1950
A metal 'van' top was made available as an alternative to the normal canvas roof, and at the same time the freewheel feature was discontinued.

1952
From the beginning of the year the engine was enlarged to 1,997cc with a larger cylinder bore.

1954
Wheelbase of the original version was increased to 86in, with a 9in increase in overall length. At the same time a long-wheelbase 107in version was also made available.

October 1956
Each basic version was given an extra 2in in the wheelbase – to 88in and 109in respectively. These have been the standard dimensions ever since. The 107in station wagon was, however, continued until September 1958.

June 1957
A diesel engine option was offered for the first time. The engine was 2,052cc, with overhead valves, and entirely different from the 2-litre petrol engine. It became available on short- and longer-wheelbase vehicles.

April 1958
The Series II Land-Rover replaced all existing vehicles except for the 107in station wagon. Styling features included 'barreled' sides. Original petrol engine was discontinued in favour of a new overhead valve 2,286cc petrol engine sharing much of the diesel's engineering and production tooling. Wheel tracks were up, and turning circles reduced.

September 1961
Land-Rovers became Series IIA, when the diesel engine was enlarged to 2,286cc, identical with the petrol engine's capacity.

September 1962
A forward-control Land-Rover was announced, using 75 per cent of existing components. Based on the existing 109in wheelbase chassis, it had a new overframe, raised and much altered cab, and a 30cwt payload (25cwt across country). Only the 2,286cc petrol engine version was offered.

September 1966
Wheelbase of the forward-control machine advanced to 110in, and the following engines were offered – 2,286cc petrol, 2,286cc diesel, and 2,625cc six-cylinder petrol. The four-cylinder petrol engine was not available on the British market. Wheel tracks were increased by 4in.

135

April 1967
The 2,625cc six-cylinder petrol engine became available on 109in normal-control Land-Rovers.

Summer 1967
A 'luxury pack' of three deeply padded front seats became optional on normal-control Land-Rovers.

Spring 1968
To satisfy certain new legal requirements, the headlamps were moved out from the grille panel to a wing position, the first obvious change in styling since the vehicle's introduction 20 years earlier.

September 1968
A special military version of the 88in Land-Rover was unveiled. Called the '½-ton', which denoted its payload, it used standard chassis parts but a lightweight body, much of which could speedily be removed for transportation by aircraft or by helicopter.

June 1970
Launch of the entirely new Range Rover, related to the Land-Rover but mechanically very different. The Range Rover had a 100in wheelbase, the light-alloy 3,528cc vee-8 engine, permanent four-wheel drive, and an attractive estate car body. It was altogether more of an 'up-market' vehicle.

October 1971
Series III Land-Rovers replaced all existing normal-control versions. These were mechanically similar to Series IIA vehicles, but with a new all-syncromesh gearbox, full-width facia styling, and a new front grille. The 88in and 109in wheelbases, with the choice of three engines, was continued.

September 1972
Debut at the Commercial Vehicles show of the latest forward-control Land-Rover – the 101in machine. Production did not begin until 1975, and the machine was supplied only to the British armed forces. It used the Range Rover's vee-8 engine and permanent four-wheel drive transmissions, with a brand new chassis and a simple forward-control pick-up body. Rear power take-off drive allowed the latest Rubery Owen powered trailer to be added.

1976
Production and delivery of the millionth Land-Rover.

APPENDIX 3

LAND-ROVER AND RANGE ROVER PERFORMANCE
(From *Autocar* and *Motor* road tests)

Land-Rovers

	Series I *86in WB* *7-seat* *station* *wagon* *1,997cc* *petrol* Autocar test, *4/3/55*	*Series I* *107in WB* *10-seat* *station* *wagon* *1,997cc* *petrol* Motor test, *18/7/56*	*Series IIA* *88in WB* *pick-up* *2,286cc* *petrol* Autocar test, *19/11/65*	*Series IIA* *109in WB* *12-seat* *station* *wagon* *2,625cc* *petrol* Autocar test, *13/7/67*
Maximum speed (mph)	58	58	67	73
Acceleration through gears (secs)				
0–30mph	7·1	7·8	6·9	5·8
0–40mph	—	14·7	12·3	11·3
0–50mph	25·5	28·9	19·9	17·1
0–60mph	—	—	36·1	29·0
0–70mph	—	—	—	—
0–80mph	—	—	—	—
0–90mph	—	—	—	—
Standing start – ¼-mile (secs)	25·7	26·2	24·0	23·6
Top gear acceleration (secs)				
10–30mph	11·4	12·7	10·2	11·3
20–40mph	12·6	15·7	13·8	13·1
30–50mph	18·2	24·7	18·2	15·8
40–60mph	—	—	22·8	19·5
50–70mph	—	—	—	—
60–80mph	—	—	—	—
70–90mph	—	—	—	—
Fuel consumption (mpg)				
overall recorded	21·0	18·2	18·3	13·8
typical (all conditions)	24	21	20	15
Weight (lb)	2,968	3,444	3,010	3,948

Range Rovers

Series III 109in WB truck cab	Series III 88in WB pick-up				
		vee-8	vee-8	vee-8	vee-8
2,625cc petrol	2,286cc petrol	3,528cc petrol	3,528cc petrol	3,528cc petrol	3,528cc petrol
Autocar test,	Autocar test,	Autocar test,	Motor test,	Motor test,	Autocar test,
28/10/71	18/1/73	12/11/70	16/1/71	25/1/75	31/10/75
69	68	91	99	95	99
6·6	5·8	4·3	4·2	4·5	4·7
11·3	10·9	6·3	6·2	6·7	7·0
17·0	16·8	10·0	9·3	11·2	10·8
31·7	29·1	13·9	12·9	15·0	14·6
—	—	18·6	17·7	20·2	20·3
—	—	28·1	25·6	30·4	30·7
—	—	42·4	36·6	45·9	47·1
22·9	22·6	19·1	18·7	19·5	20·1
11·9	9·7	10·2	—	—	11·6
12·1	9·9	9·1	9·1	10·2	11·0
14·4	12·1	9·6	9·1	10·2	10·5
24·5	19·2	9·8	9·4	11·1	11·4
—	—	10·6	10·6	12·3	13·6
—	—	14·2	13·6	15·7	16·8
—	—	24·4	20·2	—	26·8
14·9	18·0	14·4	14·8	13·9	14·1
15	18	16	18	16	16
3,582	3,090	3,880	3,864	3,842	3,861

APPENDIX 4

SALES PER FINANCIAL YEAR
1947-8 to 1974-5

Land-Rover

	Annual sales	Cumulative sales		Annual sales	Cumulative sales
1947-8	48	48	1961-2	37,139	346,293
1948-9	8,000	8,048	1962-3	34,304	380,597
1949-50	16,085	24,133	1963-4	42,569	423,166
1950-1	17,360	41,493	1964-5	45,790	468,956
1951-2	19,591	61,084	1965-6	47,941	516,897
1952-3	18,570	79,654	1966-7	44,191	561,088
1953-4	20,135	99,789	1967-8	44,928	606,016
1954-5	28,882	128,671	1968-9	50,561	656,577
1955-6	28,365	157,036	1969-70	47,538	704,115
1956-7	25,775	182,811	1970-1	56,663	760,778
1957-8	28,656	211,467	1971-2	52,445	813,223
1958-9	28,371	239,838	1972-3	49,724	862,947
1959-60	34,168	274,006	1973-4	45,169	908,116
1960-1	35,148	309,154	1974-5*	54,298	962,414

The above figures include all Land-Rovers shipped abroad as CKD packs, but not those built by MSA in Spain.

* In 1975 the financial year ended on 26 September.

The first Land-Rover was delivered in July 1948 after extensive prototype trials. The quarter-million mark was reached in November 1959, more than eleven years after the vehicle's launch. The half-millionth Land-Rover was built in April 1966, the second quarter-million being built in 6½ years. The 750,000th Land-Rover was built in June 1971, a little over 5 years after the half-millionth. The millionth Land-Rover was built in May/June 1976, and sales show no signs of falling.

Range Rover

	Annual sales	Cumulative sales		Annual sales	Cumulative sales
1969-70	86	86	1972-3	6,519	14,652
1970-1	2,537	2,623	1973-4	8,604	23,256
1971-2	5,510	8,133	1974-5	10,516	33,772

APPENDIX 5

SALES – HOME AND ABROAD

The Land-Rover was intended for export, and almost three-quarters of them have gone abroad. Sales figures to the end of the company's financial year on 30 September 1975 are as follows:

Total number of Land-Rovers built	962,000
Home sales (18 per cent)	176,000
Sales to British government (8 per cent)	74,000
Export sales (74 per cent)	712,000

For the Range Rover, the comparative figures are the following:

Total number of Range Rovers built	33,500
Home sales (36 per cent)	12,000
Export sales (64 per cent)	21,500

Land-Rovers have sold all over the world. Table 3 gives the twenty countries buying the most for each calendar year from 1964 to 1974 inclusive.

TABLE 3

Land-Rover exports – best twenty countries

1964		1965		1966	
Australia	4,161	Australia	3,099	Australia	2,685
S. Africa	1,839	S. Africa	2,371	Fr W. & EqA	1,695
Fr W. & Eq A (Fr W.		USA	1,840	S. Africa	1,462
& Eq Africa)	1,159	Fr W. & Eq A	1,392	PWA	1,304
East Africa	1,116	Malaysia	1,267	Iran	1,252
Switzerland	1,012	New Zealand	1,125	New Zealand	1,220
USA	952	East Africa	1,117	USA	1,137
New Zealand	919	Zambia	980	Libya	1,136
PWA (Port W. Africa)	891	Switzerland	912	Thailand	1,061
Malaysia	858	PWA	837	Zambia	1,037
Persian Gulf	722	Rhodesia	706	Tanzania	947

1964		1965		1966	
Thailand	694	Persian Gulf	680	Nigeria	912
Chile	669	Jordan	661	Persian Gulf	795
S. Rhodesia	655	Thailand	653	East Africa	759
Nigeria	654	Germany	645	Saudi Arabia	589
Libya	599	Algeria	613	Malaysia	586
Saudi Arabia	525	Tanzania	589	Canada	503
Algeria	519	China	575	Sudan	458
Persia (Iran)	477	Nigeria	574	Switzerland	450
Aden (inc Yemen)	428	Libya	552	Iraq	416
Lebanon	377				

1967		1968		1969	
Australia	4,166	S. Africa	3,638	Australia	5,005
S. Africa	2,323	Australia	3,600	S. Africa	2,327
Zambia	1,426	Fr W. & Eq A	1,709	Persian Gulf	2,026
Tanzania	1,309	Persian Gulf	1,523	Fr W. & Eq A	1,907
Nigeria	1,129	Nigeria	1,416	Tanzania	1,426
Persian Gulf	1,122	Tanzania	1,300	Iran	1,423
Thailand	1,096	Zambia	1,200	Nigeria	1,272
CNF	1,093	Libya	1,016	USA	1,222
PWA	941	Thailand	878	Malaysia	1,218
East Africa	794	Angola (ex-PWA)	822	Angola	1,052
New Zealand	715	Congo	790	Libya	1,046
Libya	686	Malaysia	726	Zambia	981
Switzerland	649	Jugoslavia	659	New Zealand	888
Malaysia	573	Morocco	626	Thailand	836
Port E. Africa	495	Saudi Arabia	616	Indonesia	790
Sudan	481	Ghana	612	Sudan	756
Jugoslavia	458	Algeria	553	Jugoslavia	726
Rep Cameroon	422	Kenya	550	Costa Rica	692
USA	415	Switzerland	544	Switzerland	655
Ethiopia	365	Iran	542	Algeria	558

1970		1971		1972	
S. Africa	3,320	S. Africa	3,153	Iran	3,051
Australia	3,296	Australia	3,060	Australia	2,826
Tanzania	1,618	Zambia	2,902	Nigeria	1,707
Iran	1,573	Iran	2,188	Malaysia	1,573
Malaysia	1,440	Tanzania	1,981	S. Africa	1,289
New Zealand	1,382	Nigeria	1,899	Tanzania	1,276
Nigeria	1,363	Malaysia	1,436	Libya	1,146
Zambia	1,301	New Zealand	1,249	USA	1,114
Kinshasa	1,110	Kinshasa	990	Switzerland	844
Angola	1,056	Switzerland	906	Angola	794
USA	873	Angola	826	Algeria	782
Switzerland	861	USA	756	Dubai	738
Costa Rica	798	Cameroon	724	Sudan	716
Venezuela	588	Ghana	720	Portugal	691

1970		*1971*		*1972*	
Thailand	570	Costa Rica	673	Singapore	676
Libya	531	Libya	613	Costa Rica	596
Cameroon	516	Portugal	609	Italy	594
Morocco	502	Abu Dhabi	598	Syria	558
France	454	Singapore	562	France	545
Mozambique	452	Thailand	547	Mozambique	527

1973		*1974**	
Australia	3,139	Australia	2,400
Iran	2,887	Iran	2,256
S. Africa	1,763	S. Africa	2,164
Nigeria	1,501	Turkey	1,234
USA	1,246	Nigeria	1,003
Turkey	957	Zaire	896
Switzerland	925	Italy	866
Libya	912	Zambia	849
Tanzania	871	Libya	839
Malaysia	785	Switzerland	825
Zambia	772	Tanzania	818
Angola	708	Belgium/Lux	788
France	680	France	780
Mozambique	668	Kenya	725
Kenya	648	Dubai	690
New Zealand	633	Mozambique	601
Costa Rica	609	Norway	594
Thailand	602	Malaysia	577
Algeria	572	Muscat & Oman	555
Muscat & Oman	548	Lebanon	504

* The Land-Rover was withdrawn from the United States market during 1974, because of the very high potential cost of tooling to meet North American safety and exhaust emission regulations.

ACKNOWLEDGEMENTS

No author could possibly attempt to write the story of the Land-Rover without a lot of help from the Rover company. I asked numerous questions, begged many favours, and demanded some off-beat statistics. No one complained – everybody helped. I am particularly indebted to David Bache, Stan Banting, Gordon Bashford, Tom Barton, Russell Brookes, John Carpenter, Nick Carter, David Crewdson, Richard Foster, James Gee, Eric Hawkins, Bernard Jackman, Keith Kent, Alan Luckett, Alec Mackie, William Martin-Hurst, Bill Mason, Harry Mills, Dick Oxley, Lt-Col Pender-Cudlip, Tony Poole, Dick Richter, Bernard Smith and Jack Swaine.

I also thank the following for their help: Ray Hutton and Warren Allport (Editor and Assistant Editor, *Autocar*), Roger Bell and Hamish Cardno (Editor and Deputy Editor, *Motor*), Ted Fellows (Editor, *Power Farming*), Gethin Bradley, and the Public Relations Department of the Ministry of Defence.

INDEX

Numbers in italic refer to illustrations

Fairye winches, 82
Ferguson, Harry (and products),
    19, 38, 128–30
Ford, Model T, 66
Forestry Commission, 71
FVDE (renamed from MVEE
    and FVEE), 49–56

Garnier, Peter, 121
General Motors (and Buick), 77–
    8, 83, 115–16
George, HM the King, 38
Girling, 103
GKN, 103
Goddard, Lord Chief Justice, 32
Guide to Land-Rover Expeditions,
    64–5

Haflinger, 130
Hillman, William 128
Humber-Hillman company, 15
Hundred Days of Darien, 67

Issigonis, Alec, 49

Jackman, Bernard, *17*
Jaguar, 116, 132
Jeep, 12, 19–25, 28–9, 39, 44,
    49, 51, 53, 55, 113, 129

Keikhaefer, Carl, 115
Kidson, C., 32
King, Spencer, 84, 113, 114, 116

Laycock overdrive, 82, 112
Leyland Motors, 118, 131
Light Gun (105mm), 56
Lucas, 103

Mackie, Alec, 100
Marples, Ernest, *36*

Martin-Hurst, William, 77, 115,
    116, 131, *17*
Massey-Harris company, 129
Mason, Bill, 100–2
Mercedes, 115
Mercury Marine, 115
Midland Motor Cylinder Co, 103
Morgan car, 116
Motor magazine, 137–8

NATO, 53
North Vietnam, 89
Nuffield Mechanisations Ltd
    (and FV1800/Champ pro-
    ject), 49–51, 74

Oxley, Dick, 111

Pender-Cudlip, Lt-Col, 51
Pogmore, Col, 13
Poole, Tony, 80, 84
Poppe, Olaf, 24, 96
Proteus-Bluebird project, 71

Qualcast Foundries, 103

RAC Dewar Trophy, 122
Rollert, Ed, 116
Rolls-Royce B40 engine, 49
Rolls-Royce Meteor engine, 12
Road Research Laboratory, 71
Rover
    premises: Acocks Green, 12,
    15, 92, 93, 99; Cardiff, 40,
    93; Chesford Grange, 12,
    106; Clay Lane, 91, 92, 99;
    Coventry, 15, 92; Garrison
    Street, 93, 95, 96; Helen
    Street, Coventry, 12, 16, 91,
    92; Percy Road, 40, 93, 99;
    Perry Barr, 40, 93, 99;
    Solihull, 12, 14, 15, 16, 20,

INDEX